P9-CCJ-618

The Carbon Footprint Wars

The Carbon Footprint Wars

What Might Happen If We Retreat From Globalization?

STUART SIM

EDINBURGH UNIVERSITY PRESS

Edinburgh University Press Ltd
22 George Square, Edinburgh

www.euppublishing.com

Typeset in 10.5/13 pt Palatino
by Servis Filmsetting Ltd, Stockport, Cheshire, and
printed and bound in Great Britain by
CPI Antony Rowe, Chippenham and Eastbourne

A CIP record for this book is available from the British Library

ISBN 978 0 7486 3766 9 (hardback)

Contents

Acknowledgements

My thanks go to Jackie Jones at Edinburgh University Press for all the sensitive and invariably sensible editorial help received over the course of this project. Wendy Lee was a swift and efficient copy editor; James Dale equally efficient in organizing the compilation of the final text. Dr Helene Brandon played a critical part in developing the original idea, and was, as ever, extremely supportive throughout the whole writing period.

Preface

Worried by global warming? The bulk of the evidence on climate change certainly suggests you should be. The world's carbon dioxide levels are at their highest for 650,000 years, rising more rapidly than expected and pulling the average global temperatures up with them as they go.[1] To continue in this way is to threaten the very basis of our civilization. But you should also be worried about what might be done in the name of arresting global warming's growth. If some of the more radical steps being suggested to deal with climate change were put into practice, we could find ourselves in the middle of a global socio-economic disaster. It could mean the West winding down globalization as we currently know it, with all the horrendous consequences this would have for a developing world that is deeply dependent on the system, unfair though it generally is to those countries in its present form. Climate change would have claimed yet another set of victims, and the world politically would never be quite the same again. The climate change debate is full of projected solutions, and those solutions all have their consequences – but they might not always prove to be the consequences we wanted, or thought would be produced when we undertook a particular course of action.

So this book is in the nature of a thought experiment: what might happen – politically, socially, and economically – if we retreated from globalization in order to counter global warming and the threat of what has been called 'ecocide', the wanton destruction of our environment?[2] The conclusions drawn should give us pause for serious thought: serious enough to speculate on how we might go about preventing ourselves from ever reaching the point where we have to contemplate taking the radical steps being mooted by some.

I am neither a scientist nor an economist, and global warming is a topic which tends to be dominated by the views of scientists – including science writers – and economists, representatives of disciplines that can often be impenetrable to outsiders. Even when politicians become involved, they are generally basing their pronouncements on the findings of those two groups. Al Gore's film *An Inconvenient Truth*, for example, was heavily reliant on the work of the controversial American scientist James Hansen;[3] the British government commissioned *The Stern Review* to outline the economics of global warming so as to inform its own policy on the issue.[4] But this is not a debate that can be left mainly to the scientists and economists; the 'carbon footprint wars', as I am going to call them, involve us all, and they need to be opened out to a wider range of perspectives, as they will be in this study.

Part I

The Problems

1
Introduction
The Carbon Footprint Wars: What is at Stake?

Global warming is the outstanding challenge facing the human race at present, and radically reducing our carbon footprint is its most pressing requirement. There may be disagreement as to what is causing the warming, or how we should go about addressing it, but only a few die-hard sceptics are resisting the idea that the process actually is occurring and that the evidence is all around us inviting analysis and interpretation. And those die-hard sceptics are in many cases representing entrenched interests, such as the international oil companies for whom increased carbon emissions mean increased profits (and their profits have been hitting record heights of late), so their opposition has to be responded to with a certain amount of scepticism. But are we all fully aware of the political and sociological complexities of taking the challenge really seriously? Or of the range of consequences that might ensue from vigorously pursuing a significant reduction (bearing in mind that time is not on our side, and that a low-level programme of gradual change is unlikely to be sufficient)? This study argues no on both counts, also insisting that there are dangers inherent in many of the projected solutions – such as retreating from the spread of globalization, the current paradigm for world trade.

How we respond to these dangers will dictate how successful we are likely to be in maintaining a stable world order based on mutual respect for cultural difference – as well as for the global environment. The war of words being waged over the appropriate way to deal with our collective carbon footprint has critical implications for us all, which I will examine in detail – while in no way disputing the fact of global warming or the threat it clearly poses to all of us and our way of life. Or, for that matter, taking an uncritical line on globalization as it is practised, which most certainly leaves a great deal to be desired and is ripe for change.

The scientific press is full of dire warnings as to our collective fate if we fail to effect a very substantial reduction in carbon emissions, and to do so quickly. We are now deemed to be living in the anthropogenic era, where human activity is the critical shaping force on the environment rather than the Earth's physical systems themselves, so the onus is firmly on us to take the appropriate action; only we can save ourselves. Even just a few years ago this was thought to be a problem for future generations to face, thus breeding a certain complacency, but such projections have been drastically foreshortened and the problem is manifestly here now. As the climate historian and archaeologist Brian Fagan has warned us, we must never forget that climate change can catch us unawares, being essentially 'unpredictable and sometimes vicious' when it strikes.[1]

No-one has been more forceful in making the point that we have allowed ourselves to be caught unawares than James Lovelock. In his apocalyptic vision of the future, the process of climate change is so far advanced that we should already be planning for basic survival in what is about to become a very inhospitable environment.[2] Lovelock's rhetoric is extremely emotive. He thinks that the Earth – which for him is a complex, sensitively balanced system known as Gaia – is sick, and that it will take possibly as long as 100,000 years to recover to a state of health: 'We have given Gaia a fever and soon her condition will worsen to a state like a coma.'[3] The human race, a constituent part of Gaia, will have to suffer through this period, when much of the Earth will be uninhabitable, as best it can, even if it means being reduced to living only in the Arctic, the rest of the planet having become a no-go area. We are the culprits and we shall have to pay the price for our recklessness in the use of the Earth's resources over the last century or so. Humanity will survive, but with only a fraction of our present numbers. Not everyone is as pessimistic as this (George Monbiot's recent book offers a guide as to 'How We Can Stop the Planet Burning', as a case in point[4]), but Lovelock's views nevertheless carry a great deal of weight. It is to be hoped that the situation is not as drastic as he claims, but worst-case scenarios of this kind are at the very least thought-provoking. 'We're not getting it,' in the frustrated words of the science writer Bill McKibben, and we must be made to understand where our inertia over the issue could lead us.[5]

Perhaps one of the reasons why 'we're not getting it' is that the carbon footprint wars are full of what the philosopher Jean-François Lyotard has called 'differends':[6] that is, incommensurable world-views, where disputants are simply not on the same wavelength in terms of what they are talking about or trying to achieve. Enthusiastic supporters of globalization see the world almost exclusively in economic terms, where economic growth – no matter how unequally spread amongst nations it might be – is a sign of success; for those concerned about global warming, economic growth is a source of more carbon emissions, so a sign of failure. Developing renewable energy sources – solar or wind power, for example – is the answer to our problems with fossil fuel for some thinkers, whereas to others it is a potential destroyer of complex ecosystems with unknown consequences for our world. Various companies can see only profit when engaged in deforestation programmes, opponents only a reduction in the Earth's ability to store carbon that amounts to a criminal act on the part of the deforesters. Differends like those (and there are many others to report) have to be overcome if there is to be any real progress made in stabilizing global temperatures, and I will be paying close attention to the impact they are having on debate over the course of this book.

Kyoto and After

The Kyoto Treaty (negotiated 1997, in full operation from 2005) set targets for all nations as to carbon emissions, recommending that the world's industrialized nations achieve an average reduction of 5.2 per cent by 2012 compared to their 1990 levels.[7] In real terms, this meant a cut of 29% on what the levels would have been expected to be by that point, given the steady increases that have taken place year by year since 1990. But as yet there is no overall global policy on how to realize these targets – or even to abide by them at all, with the USA, the world's largest economy, pointedly opting out, thus rendering the exercise little more than cosmetic (other notable non-signers were Australia and Kazakhstan). President George W. Bush preferred the idea of voluntary controls, with each country retaining the right to decide what would best suit its economy and lifestyle, and in his view committing to Kyoto, as he wrote to some senators, 'would cause serious harm to the US economy'.[8] Not surprisingly, this has proved to be a recipe for stasis. Most commentators concede that little of real significance is likely to occur in this area before the Kyoto protocols run out in 2012, at which point the whole deal will have to be renegotiated (it is also worth noting that no nation was put under any obligation to reduce air traffic emissions, which surely has to be reconsidered next time around).

Nationalist aspirations only too easily get in the way of international consensus on this issue, however, with the explosive growth of both the Chinese and the Indian economies in recent years constituting a particular source of worry for the immediate future. They may be starting from a low base, but with populations in excess of a billion each they have the capacity to make an enormous impact on global pollution. Industrial and technological expansion on this scale can only exacerbate the problem that the West has been in the forefront of creating to date – which makes our criticisms of Indian and Chinese economic policy ring more than somewhat hollow. Even if America were to fall into line and meet the Kyoto Treaty's recommendations, the problem may just move elsewhere. Lovelock's pessimism about our collective fate can come to seem all too plausible at such points.

Nevertheless, there is general agreement as to where action is urgently needed if our carbon footprint is to be reduced significantly on a longer-term basis. Air travel is one of the highest-profile targets, as is car travel. The consumer society in general is wasteful of resources, with luxury goods and foodstuffs being ferried around the globe, using up vast quantities of fossil fuel to reach their ultimate destination in the shopping complexes of the West. Food miles have become a topic of increasing debate in recent years, with many Third World countries exporting fruit and vegetables to the West so that there is a year-round supply in supermarkets of hitherto only seasonally available products. Disengaging ourselves from both mass tourism and the consumption of exotic or locally out-of-season food will be no easy task; the West is seriously addicted to both, which are now firmly engrained into our lifestyle, for all the feelings of guilt they can induce in many of us on occasion. It may be a pleasant experience to have products like asparagus and strawberries on our tables in December and January in northern Europe, but they come at a carbon cost that has to be acknowledged; baby vegetables are emission-loaded.

Addressing Climate Change

We are still in the very early stages of addressing this problem, and most of us are still confused as to how to react, at either a private or public level; but a range of measures have been suggested – some of them extreme, such as tinkering with the composition of the Earth's biosphere or making air travel, and air freight, prohibitively expensive to deter users. Until there is a large-scale disaster in the West that is unmistakably the product of global warming (a massive increase in destructive storm activity claiming many lives, for example, or enough of a rise in sea levels to flood the coastal plains where much of the population currently lives[9]), then the more extreme responses are unlikely to be put into practice. What is likely, however, is that we will move incrementally towards these as the effects of global warming accelerate – as most of the scientific community argue is already well under way, and quite possibly irreversibly so in many cases. A measure that strikes us as extreme now may not seem so once we have passed some critical tipping point, such as a disintegration of the Antarctic ice sheets, which would trigger a phenomenal rise in sea levels (50 metres plus). From that point onwards, damage limitation would be more of a priority for the political class than maintaining an expansionist-minded world trade system. The days of carbon offsetting, increasingly coming to be regarded as a dubious activity anyway (some recent studies suggesting that trees cannot absorb as much carbon as was previously thought, for example[10]), will be long gone.

Severe restrictions on tourism to reduce air travel, on food imports to reduce air miles, on the spread of globalization – all of these are possible in the near future, therefore, if current climate trends continue unabated. Cheap package holidays could well become a thing of the past – as could the availability of foods out of season, perhaps exotic foods in general. Engineering the biosphere to be cooler, or geoengineering as it has come to known, is a less probable occurrence, given the enormous cost and high risk factor – although it cannot be ruled out altogether. Again, it will be a question of how serious the situation has become and thus conducive to desperate remedies; faced with tinkering with the biosphere or the mass extinction of humanity, we can well imagine the likely choice to be made. Having said that, previous attempts at altering the course of nature on any large scale are not exactly encouraging. As the sociologist Zygmunt Bauman has noted of such grandiose schemes undertaken by the Soviet regime in the heyday of communism:

> Deserts were irrigated (but they turned into salinated bogs); marshlands were dried (but they turned into deserts); massive gas-pipes criss-crossed the land to remedy nature's whims in distributing its resources (but they kept exploding with a force unequalled by the natural disasters of yore) . . . Raped and crippled, nature failed to deliver the riches one hoped it would; the total scale of design only made the devastation total.[11]

If geoengineering ever does become a method of dealing with global warming, then we have to hope that in this case nature does actually manage to deliver; failure at this scale would dwarf any of the after-effects suffered through Soviet environmental engineering, bad though these were.

Such measures will not be popular, but disaster will no doubt help pave the way for their acceptance, grudging though that may be in the first instance. Yet in attempting to resolve, or at the very least to stabilize, the problem of global warming, might these initiatives have unintended consequences that could be socially and politically very damaging in their turn? Many Third World countries rely heavily, unhealthily so given the vagaries of the market, on tourism and food exports, and in both cases air transport is a critical, if not indispensable, element of delivery. When it comes to fresh food transported over long distances, speed is of the essence and air freight the logical method to employ. Neither would holiday journeys of several hundred or thousand miles be feasible on a mass basis unless by plane. Some such economies would most probably collapse given any really savage restriction on air travel. Brazilian rain forest is being systematically destroyed to create grazing pasture for the cattle industry which then produces meat for such multinational corporations as McDonalds, and this accelerates globing warming too. The Amazon basin's rain forest plays a critical role in the Earth's ecosystem, and its progressive reduction is a cause of considerable concern. Curtail this process significantly, however, and a huge work-force, much of it already living at something like subsistence level, will lose their livelihood – with little real hope of any replacement employment coming to the rescue. (Admittedly, there are various complicating factors in such an argument, with the journalist George Monbiot pointing out that those who work in the Amazon beef-producing industry are often in a condition bordering on slavery.[12] This is an unacceptable practice and should be campaigned against vigorously. But it is a measure of the plight of much of the Third World's population that the alternative, unemployment in a largely welfare-free environment, can be even worse – this being a classic example of what has been called a 'tragic choice'.[13])

McDonalds cannot, and should not, be seen as the saviour of the Third World; but simply removing corporations such as this from the scene altogether is not the answer to the Third World's problems either. Establishing a trade-off between economic survival and ecological disaster is going to become one of the major geopolitical problems of our times. This has been recognized by the creation of the Coalition for Rainforest Nations, an organization which is campaigning vigorously for cash incentives to be offered to poorer nations if they agree to conserve rather than cut down their forests as cash crops. World deforestation has been growing at a worrying rate, and it is clear that unless this trend is reversed climate change will be accelerated considerably. There is scepticism amongst many other climate campaigners as to whether this idea will work, however, with political corruption being cited as a likely barrier to its success, but such trade-offs have to be explored if any progress at all is to be made.[14]

Fear is already beginning to be expressed that failure to find a workable trade-off could trigger wars over scarce resources, with the peace group International Alert, in their 2007 report *A Climate of Conflict*, identifying Africa, Asia, and South America as the areas most at risk. The scale of the problem the report outlines is alarming:

> There are 46 countries – home to 2.7 billion people – in which the effects of climate change interacting with economic, social and political problems will create a high risk of violent conflict. There is a second group of 56 countries where the institutions of government will have great difficulty in taking the strain of climate change on top of all their other current challenges. In these countries, though the risk of armed conflict may not be so immediate, the interaction of climate change and other factors creates a high risk of political instability, with potential violent conflict a distinct risk in the longer term. These 56 countries are home to 1.2 billion people.[15]

This total of 3.9 billion people constitutes nearly 60 per cent of the world's population, so this is not a situation that can be ignored. Soon we shall all find ourselves being affected in some way or other by what the report's authors neatly dub the '*consequences of consequences* of climate change' in the less developed parts of the world.[16] They cite Darfur as an 'exemplary case' of those consequences of consequences.[17] There, the combination of persistent drought and long-standing disputes over the always scarce resources of a largely desert area has led to further strains on the environment, as warring militias seek to secure their own positions at the expense of each other in what has become a protracted and brutal conflict. The result has been further desertification of the area, thus a reduction of the available resources that can only serve to exacerbate the conflict itself. In a world where, as the report notes, water shortage currently affects 430 million people, it is not difficult to envisage such conflicts becoming increasingly common as nations struggle to keep themselves together as economically viable entities.

The collapse of Third World economies would undoubtedly create even greater pressure to migrate on the part of their populations, and the West is already – rightly or wrongly, depending on your political position – seriously concerned about its ability to withstand immigration on a far smaller scale than any such mass exodus would create. A fortress mentality could well develop, which could seriously disrupt global relations. Its foundations are already there in countries like the UK and France, where the popular press is only too happy to play the nationalist card. Sad as it is to note, for all the official commitment to multiculturalism in most Western nations there is nevertheless a substantial market for racism and prejudice, especially if it is pitched as a plea for 'our way of life' to be protected. Immigrants can find themselves being blamed for a country's socio-economic problems, and unable to do much to protect themselves from such a charge.

To scale down mass tourism – detrimental though it can be to traditional societies in a host of ways – would hardly help the cause of multiculturalism either. The less contact there is, the more alien other cultures come to seem, then the greater the chance that prejudice, both national and racial, will increase from what already is a worrying enough level. Western nations would be all too prone to withdraw into themselves (as some do periodically even now), becoming progressively, even aggressively, more monocultural – which predictably enough would be welcomed by the more reactionary forces in those societies, thus presenting a threat to the liberalizing ethos that has prevailed in modern times. (The UK government is already expressing concern about a projected population increase of 10 million in the next quarter-century, and the need for tighter immigration controls has been the reflex response to the perceived problem – with the official parliamentary opposition being, if anything, even more enthusiastic about implementing these measures.)

Globalization as it is currently practised is exploitative of the Third World, often horrendously and indefensibly so, as commentators like Naomi Klein have been at pains to demonstrate, but if it were cut back on radically then, again, many Third World economies would go to the wall, having no other resources to fall back upon to sustain themselves.[18] Even developed nations would run into difficulty if global trade were to be reduced; countries like Australia and New Zealand would struggle to exist without their export trade, lacking enough of a home market to keep themselves prosperous on the standard Western model. Trade is their umbilical cord to the West, where their socio-political roots still largely lie. We could live without New Zealand lamb in the West, but could New Zealand live without exporting so much of it?

Clearly, action to reduce our carbon footprint is not necessarily as straightforward a procedure as it may seem; unintended consequences could have almost as devastating an effect on the lifestyle of large sections of the global population as global warming is projected to do. The recent enthusiasm for the production of biofuels is another striking case in point. The massive increase in production of ethanol in the West in recent years is pushing up food prices globally, as crops such as maize are converted into fuel for cars rather than, as previously, sold as foodstuffs. In theory, this seems an idea well worth pursuing; when ethanol is mixed in with petrol, it causes greater combustion of the petrol itself, thus cutting toxic emissions from cars. A shift over to ethanol-based fuel could make a real difference in the size of our carbon footprint, but as food crops are cut back on, so food prices inevitably rise, the supply and demand equation of market economics asserting itself. As one would expect, that has a far greater impact on the Third World than on the West's developed economies. The former have far less margin for error, fewer reserves to draw upon, when world markets experience dramatic shifts of emphasis, as in this instance. A modest increase in the price of a staple item in the West, irritating though it can be to the average consumer, can be absorbed; in the Third World, on the other hand, it can represent the difference between survival and starvation. Neither is it even clear any longer that biofuel will serve to reduce carbon emissions; as we shall go on to discuss in Chapter 9, some commentators have in fact started to argue the opposite.

This should not be construed as an argument to jettison all attempts at curbing global warming and continue as we are doing, but rather to identify and then factor in the threat of unintended consequences to the process as much as possible. There is a very real danger that the West will act to protect itself at the expense of the Third World, and this is surely to be resisted on humanitarian grounds. A first step is to speculate on what those unintended consequences may turn out to be, and to consider how they might be addressed. It is a useful exercise to outline the worst-case scenarios in order to make us think through the dangers that exist in any radical approach to arresting the course of global warming. There are political as well as technical considerations that must always be borne in mind, as well as moral ones with regard to our mutual responsibilities to all the globe's cultures; we are all in this together and cannot regard any part of the world's population as expendable in the name of some assumed greater human good (one can already imagine how such arguments would be packaged to gain public acceptance in the West). This book will address these considerations by spelling out what the problems are (Part I), and by looking at a wide range of the projected solutions (including those involving geoengineering, which are beginning to exercise the scientific press, for all their rather far-fetched nature (Part II)). It will then examine the potential unintended consequences of their application, deliberately provoking argument by flagging up worst-case scenarios – economic, socio-political, technological, and environmental – with their capacity to create dystopias that would be catastrophic for the Third World in particular (Part III).

Redrawing Our Political Narratives

Dealing with these scenarios plunges us into a series of paradoxes which turn many of our traditional conceptions of politics upside down, and that will be the concern of Part IV. The left generally is opposed to the spread of globalization, seeing it as exploitative and culturally destructive, but perhaps it is one of the main things that promotes a social climate receptive to multiculturalism? The left is also ambivalent about tourism and its impact on Third World cultures, but many of these would not be viable at all as things stand, were it not for the income that is generated from mass tourism. Food miles may be bad for carbon emissions, but they are also a lifeline to many marginal economies. If we are to reduce the scale of such activities then we are surely under an obligation to provide some kind of substitute for the income they bring in; otherwise many nations will simply go under. (As a case in point, the failure to provide satisfactory substitutes for opium production in marginal economies such as Afghanistan has meant that farmers have just gone on growing the crop illegally in defiance of Western pressure. This is a case of refusing to go under on the part of the indigenous population, but when it comes to tourism and food exports that kind of response is not on offer and there is little that can be done by way of retaliation.) A position has to be articulated between ruthless, unregulated, market fundamentalism at one end of the spectrum and traditional anti-capitalist leftism at the other, and that means reassessing how we construct our political narratives. This becomes all the more imperative when hitherto standard-bearing anti-capitalist societies such as China have now embraced advanced capitalism and become part of the problem rather than, even in the deeply compromised way that it was, any kind of prospective alternative (the Soviet empire having long since passed away). If market fundamentalist-driven capitalism is all that we can look forward to, then the future looks desperate indeed on the carbon emission front.

Traditional political narratives may, in fact, present a significant barrier to dealing with the situation overall, and thus prove to be in need of large-scale overhaul. Theorists of 'radical democracy' may help to give us a lead in this respect, given their commitment to moving away from the tired old clichés of socialist politics and instead combining forces with emerging new social movements around the globe to create a new, post-Marxist narrative more concerned with finding enterprising new solutions to political problems than in maintaining theoretical purity for its own sake (as the far left in particular has been only too wont to do down through the years).[19] For theorists such as Chantal Mouffe the goal is an open-ended pluralism where a wide range of viewpoints, some diametrically opposed to each other, is continually being expressed – and accepted as the norm by all participants in the political process (what she calls 'agonism'). Mouffe draws attention to how a consensus tends to be engineered between the leading players in most Western countries, such that the more radical views are silenced; thus her complaint about 'the typical liberal perspective that envisages democracy as a competition among elites, making adversary forces invisible and reducing politics to an exchange of arguments and the negotiation of compromises'.[20] A consensus of this order might well consign the Third World to oblivion if it was thought that this would enable the West to survive global warming. If the only other option on offer is a far-left consensus stuck in an outdated mind-set and overtaken by events, then there really is an urgent need to redraft our narratives to fit a rapidly changing reality which is outstripping the scope of our theories.

Pluralism of any kind would be at risk were we to continue with the same old narratives that have marked out left and right over the course of the last century or so, on the assumption that they constitute a clear-cut binary opposition to be chosen between by individuals. The left in particular will need to rethink its entire ethos if it is to provide any meaningful contribution at all to the campaign to combat global warming; the certainties of the past can no longer be depended upon. Ulrich Beck makes a similar point when he calls for a 'new cosmopolitanism' in our politics to address what he calls 'glocal' questions: that is, 'global and local questions which do not fit into national politics'.[21] No doubt there can be left and right solutions to those glocal questions, so there will still be political debate; but they need to be in dialogue and to agree that their focus really is glocal as opposed to narrowly national or regional.

One suggested new source of income to replace Third World dependency on Western tourism and food imports indicates just how complex the problem can be, however, and that is the idea that has been touted in scientific circles of late that large-scale desert areas – such as the Sahara – can be turned into gigantic solar power farms. The power could then be sold on to the West, where the bulk of the demand exists, reducing our need to rely so heavily on fossil fuels for the production of the energy on which technologically advanced societies such as ours depend. An obvious drawback is that this would be to destroy a large part of the Earth's remaining wilderness, thus adversely affecting the environment and its ecological balance, which can be quite delicate. Similar arguments can be mounted against wind farms, often put forward as a potentially large-scale replacement for fossil fuel-derived energy. Environmental campaigners would argue that this is the wrong way to go about reducing our carbon footprint, and would be more likely to press the case for a radical restructuring of our technologically driven lifestyles, but there are compelling arguments from the other side also.

As so often in this debate, it is not always clear what the best course of action would be – as well as the permanent danger of unintended consequences unacceptable to at least some section of society to be taken into account. It is not just a case of losing an aesthetic dimension of the world's landscape, which would be hard to sustain if it was a choice between that and actual survival (although I think the arguments still need to be voiced). Rather, we simply have no way of knowing what the longer-term effects on the environment would be of such a scheme and how this would affect us. An apparently seductive short-term solution could so easily turn into yet another long-term problem to add to the many that global warming already has set us.

Sceptical Environmentalism

One of the most interesting voices raised against the more radical plans to deal with climate change is that of the Danish political scientist Bjorn Lomborg, whose book *The Skeptical Environmentalist* created considerable controversy by its opposition to the Kyoto protocols.[22] Despite his scepticism Lomborg accepts that climate change is a reality, but argues that the emphasis should be on developing different kinds of technology to deal with it piecemeal rather than expecting radical changes of lifestyle – or assuming that people can be panicked into the latter by apocalyptic rhetoric of the kind favoured by Lovelock:

> Asking people to show goodwill and change their lives is effectively a tax on good people. We need markets and social systems that make the choices for us . . . [T]he solution will come, in the main, not from carbon dioxide taxes but from smarter technologies . . . I understand the emotional satisfaction of having everyone screaming about climate change now, yet maybe that's not the best way of delivering a solution.[23]

Lomborg makes some very pertinent points, proving that, unlike most sceptics on this issue, he is no mere defender of the status quo. He is against biofuel development, for example, arguing that '[t]here is something fundamentally wrong about taking food and turning it into fuel, at great environmental cost, and at the same time driving up food prices which especially affects the poor'[24] – and it is clear that many others are coming round to this point of view. He also has a good eye for unintended consequences, pointing out how a shift to organic farming to prevent deaths from pesticides might actually increase deaths from cancer by pricing some fruit and vegetables out of people's means, thus altering their diet for the worse and making them more susceptible to other diseases. This entire area of debate proves to be rife with such counter-intuitive outcomes, and we really do need to be on the look-out for these at all times; who would have thought before the event that biofuel development would have meant higher food prices?

Lomborg can be very dismissive of the predicted effects of climate change. Speaking of the predicted rise in sea levels, for example, his conclusion is that: 'We have a failure of imagination here. We fail to realise how different the world is going to be in so many other ways a hundred years from now.'[25] Lomborg is also fairly sanguine about the economic impact of global warming: 'the total damage of global warming in the coming century will be substantial – perhaps $15 trillion – yet this will only be 0.5 per cent of total economic activity.'[26]

But the question of exactly who would bear the brunt of that cost hangs tantalizingly in the air – and the West has been showing a marked reluctance to devote any substantial amount of public funding to counteract climate change. While I agree with Lomborg that panic is not conducive to good decision making over global warming, a sense of urgency is not the same thing as panic – and the urgency may be more necessary than Lomborg is willing to admit.

I will be engaging more fully with the ideas of Lomborg in Chapters 3 and 6. Maverick and polemical though he may be – and his critics have not been slow to make the point – Lomborg's contribution to the debate goes well beyond crude scepticism, and is never less than politically thought-provoking.

Establishing Parameters

There may be no easy solution to the problems generating the carbon footprint wars, but the parameters of the debate do need to be established, differends and all, so that we can determine what is most at issue. That is what this study sets out to achieve, posing such questions as: what are the dangers of a really robust response to global warming, and what do they require us to be particularly attentive to, socially and politically? And where would such attentiveness leave our current notion of traditional political allegiances? The *Stern Review*, commissioned by the British government in 2005, provided comprehensive statistics of the likely economic costs of moving towards 'a low-carbon global economy', and argued that '[t]he costs of stabilising the climate are significant but manageable';[27] but what happens when we move beyond the economics into the murky world of politics? The *Stern Review*'s recommendations have been very controversial and will be discussed in more detail in Chapter 5, but we will start the process of answering such questions by investigating in the next two chapters the arguments for and against global warming as a physical phenomenon: what exactly is at stake in this debate?

2
Global Warming: The Evidence For

We turn now to consideration of the scientific evidence for global warming. There are various kinds to take note of, including changing global weather patterns, systematic temperature rises in recent years, increased storm activity, the melting of the Antarctic and Greenland ice shelves and ice sheets, and the opening up of the fabled Northwest Passage during the summer months in the Arctic. In the case of the latter, it is ironic that what would have thrilled our forebears is now being viewed by us with something approaching dread (most of us, anyway; it has been reported that one Canadian government minister saw this event as a welcome opportunity for the country to collect more shipping fees for travel in its territorial waters[1]). Computer modelling of current trends regarding the weather and its effect on the Arctic and Antarctic regions is not exactly encouraging. Projections suggest that sea levels could rise by several metres, for example, swamping coastal cities around the globe and perhaps even rendering entire countries uninhabitable. They also suggest that the Arctic will soon become an ice-free zone for much of the year; that extreme weather events such as hurricanes will increase in intensity; that long-term drought could lay waste vast areas of the planet's landmass, creating a full-scale agricultural crisis that would destabilize civilization. All in all, it is a depressing list to contemplate.

Computer models are just that, models, and can only forecast on the basis of the information they are fed, which is not necessarily the whole story, since we do not yet have full knowledge of the workings and interaction of all the various systems involved in the world's environment. As Fred Pearce has remarked, the problem with all such modelling 'is that it is not always easy to unpick exactly which of the elements in the model is causing the effects that you see in the printout'.[2] There is considerable scope for interpretation, and particularly so when it comes to forecasting how climate change will affect different regions of the globe, which is obviously a matter of some importance for individual countries trying to work out how they should respond to the modellers' projections.[3] This introduces an element of uncertainty into the process – sceptics might even say guesswork. As the biologist William Laurance has warned us, we have to be on our guard against '[t]he perils of trying to make linear decisions in a non-linear world'.[4] Such provisos granted, the models still tend to yield broadly similar predictions of what the impact of runaway global warming, with average temperatures soaring by 5°C or more, will be on the Earth system, and taken collectively they are distinctly alarming.

James Lovelock, Fred Pearce, Mark Lynas, George Monbiot, Gabrielle Walker, Sir David King, and Bill McKibben are among the key figures in this debate, and although all accept the fact of global warming, they present an interesting spectrum of opinion for us to ponder. Lovelock is the most apocalyptic in tone, and we shall be returning to his work at various points throughout the book as one of the highest-profile scientific commentators on the issue, whose views evoke strong responses from other scientists and the general public alike. The spectrum ranges from Lovelock's belief that we are probably past the point of no return, to the more measured response of commentators like Bjorn Lomborg and the team that compiled the *Stern Review*, who think that we can prevent the worst happening by a carefully calculated programme of activities that it is within our capabilities to engineer. Even Lovelock makes some recommendations for reducing our carbon footprint, while simultaneously warning us it may be too late for this to have any significant effect on our potentially desperate fate.

Lomborg's scepticism and Stern's economic bias suggest that they are better dealt with in Chapters 3 and 5 respectively ('Arguments Against' and 'Altering Lifestyles'), but in between these poles commentators can lurch from guarded optimism to outright pessimism and despairing predictions of doom and disaster. Perhaps the most telling comment has been that of Pearce, who admits that, ultimately, we just do not know what will happen, or what effect, for good or ill, any of our efforts to stave off the worst-case scenario will have. It is a case of hoping for the best but preparing, and in our low moments most probably fearing, for the worst, since, '[r]ight now, there is no . . . prognosis except uncertainty.'[5] Facing up squarely to the evidence piling up for global warming – year by year, study by study – it is difficult not to subside into a condition of gloom.

Revenge, the Arctic, and Niagara Falls

It is worth starting our survey of the evidence with James Lovelock, just to accustom ourselves to the most extreme vision of global warming currently on offer. Lovelock is one of the most powerful voices in what amounts to a distinctive literary genre of our time – we might call it 'apocalyptics'. It is not only distinctive, but highly popular. There is a ready audience for books, articles, films, or television programmes that proclaim we are on the verge of disaster and that civilization as we know it is probably doomed, whether from climatic change, terrorism, the overdue explosions of supervolcanoes, agricultural crisis, perhaps even from asteroid impacts; dystopian visions do sell, and they exercise a strange power over us.[6] (The latest in this line to create a stir is *A World Without Bees*, which argues that our survival would be seriously at risk if bees died out entirely. There are certainly worries being expressed about the marked decline in bee numbers, particularly in America, in recent years.[7]) A *New Scientist* editorial has suggested that we could look back to the 1972 book *The Limits to Growth*, one of the first to make use of computer modelling for predictive purposes, as the inspiration for the current crop of such visions: 'It found that if trends in population, industrialisation, pollution, food production, and resource depletion continued unchanged, resources would eventually run out.'[8] Heavily criticized as unnecessarily doom-mongering at the time (although the authors later insisted their argument 'was not about a preordained future. It was about a choice' that faced us[9]), the book now looks remarkably prescient. It has set the tone for an entire discourse, which goes as follows: because of limits to both our resources and our capabilities, the end of civilization is approaching if we go on as we are (followed up by *New Scientist* with a special feature on the topic, 'The Collapse of Civilization'[10]).

Lovelock has his own special niche in this discourse, arguing that we are heading into the worst crisis of recorded human history. Worse still, it is a crisis we are apparently powerless to halt the progress of, and humanity, or what remains of it, will just have to wait it out patiently until the Earth slowly regains a state of equilibrium. As we have seen, Lovelock's prediction is that the human race will, in a very short time – just a few generations perhaps, be reduced to a few thousand individuals, eking out a miserable life in the Arctic regions. Human history comes to seem the product of a fairly brief climatic window – in reality only about 6,000 years in duration – which is fast disappearing. We were fortunate as a species, but our luck is deserting us and we shall soon find ourselves at the mercy of natural forces totally outside our ability to control. The only move that might delay this process somewhat, in Lovelock's opinion, is a return to nuclear power, which he argues is our only relatively safe option for the generation of the energy that our advanced technological civilization insistently requires. Even that, however, will probably do no more than delay the inevitable. It is a vision of biblical harshness, with human suffering right to the fore.

Lovelock's arguments do not inspire much in the way of optimism. Complexity theorists play up the virtues of existence at what they term 'the edge of chaos', the condition where systems, such as life, are forced to make strenuous efforts to keep themselves fit and functioning.[11] Being at the edge of chaos encourages innovation and creativity to stave off the ever-present threat of collapse, and that can be an exhilarating, adrenaline-inducing situation to be in, but there is always the risk that creativity can flag or complacency set in. Some theorists in the area feel that something like this must inevitably happen with all systems, that the dynamics ultimately decline. This is especially so the more complex the systems become, with the science journalist Debora MacKenzie reporting one such team's conclusions 'that an ever-faster rate of innovation is required to keep cities growing and prevent stagnation or collapse, and in the long run this cannot be sustainable'.[12] This would appear to be Lovelock's conclusion about our current condition, that we have lost the battle this time around: 'our future is like that of the passengers on a small pleasure boat sailing quietly above Niagara Falls, not knowing that the engines are about to fail', as he rather typically and melodramatically sums it up.[13] While there are recorded cases of individuals being swept over the Falls and surviving, one would not want to bet on it.

Revenge, Continued . . .

Fred Pearce also offers a very thought-provoking prognosis as to our prospects in *The Last Generation*, which is all the more powerful for lacking the apocalyptic tone of Lovelock's work. Pearce systematically works through the evidence, giving space to the more moderate as well as the worst-case scenarios of our future, but the general tenor of his argument is decidedly on the gloomy side:

> Humanity faces a genuinely new situation. It is not an environmental crisis in the accepted sense. It is a crisis for the entire life-support system of our civilization and our species . . . In the past, if we got things wrong and wrecked our environment, we could up sticks and move somewhere else. Migration has always been one of our species' great survival strategies. Now we have nowhere else to go. No new frontier. We have only one atmosphere; only one planet.[14]

The weight of evidence suggests that things are spiralling out of control, and that it is only a matter of time before some catastrophic event shatters our social system. As a *New Scientist* article cheekily pointed out, our problems would be solved at a stroke if we had another half a planet to work with, as that is the rate at which we are using up our natural resources;[15] but failing that, we really do have to start fearing the worst. More soberly, other commentators speak of the problem of an 'overshoot' on our environmental resources, and you cannot go on doing that indefinitely.[16]

Pearce nevertheless feels we can still pull back from the brink, just as long as we put some practical measures into operation, and the sooner the better. Drawing on the work of the American academic Robert Socolow, he gives us a wish-list of these to consider. Socolow put forward a fifty-year plan based on a series of actions, which he called 'wedges', each of which had the capability of cutting annual global carbon emissions by 25 million tonnes. The effect of instituting twelve of these wedges over fifty years could be dramatic: a reduction of the projected figure of 14 billion tonnes of carbon emissions in the year 2060 to just 2 billion tonnes (the rate at present is 7.5 billion tonnes annually). If that could be done, then we might indeed have averted the worst-case scenario of runaway global warming and the probable collapse of civilization as we know it, and the proposals are all well within our technological means: a fifty-fold expansion globally of wind power and biofuels, for example, or a doubling of the world's nuclear power capacity.

While they sound within our capabilities, some of the proposals also have the potential to be environmentally very disruptive: covering an area the size of India with new forests, for example, or another the size of the state of New Jersey with solar panels. Where these areas would be on the face of the globe is left open, as is the cost. When it comes to the latter the projections are often so disparate as to produce only confusion amongst the general public. As Pearce notes, when the International Panel on Climate Change (IPCC) asked a team of economists to come up with a figure for stabilizing the atmosphere by 2100, they were given estimates ranging from $200 billion to $17 trillion. That hardly encourages global cooperation, as countries would not really know what they were committing themselves to if they did sign up for any particular programme – it would be more like a blank cheque. A figure of $200 billion would hardly tax the international community overmuch, whereas $17 trillion patently would (although it could be argued that dealing with the projected effects of global warming at a more advanced stage than now, when we could no longer avoid doing so because of adverse environmental effects, would be even more costly).

Carrying out Socolow's wedges would require a high degree of global cooperation, however, as well as a clear vision to carry them through over a period of decades, and that, sadly enough, is conspicuously lacking at present. Noting that in his capacity as a science journalist he has been warning of the growing problem of climate change for some time now, Pearce has to admit little actual progress has been made:

> Fifteen years on, the urgency of the climate crisis is much clearer, even if the story has grown a little more complicated. But we are showing no signs yet of acting on the scale necessary. The technology is still straightforward, and the economics is only easier, but we can't get the politics right.[17]

Not getting the politics right is a recurrent refrain of writers in this area (the short term invariably tending to dominate in political thinking), and the fear is that the longer we fail to do so, the more likely it is that the situation will slide irrevocably out of our control, that we shall tip over the edge into actual chaos itself. Positive though Pearce tries to be, and practical as the measures are that he puts forward with a view to keeping warming below the +2°C threshold, the overall feeling communicated by his analysis of the evidence is that we cannot really consider ourselves masters of our destiny as a species any more; that we are caught up in a process that goes well beyond our understanding and that is going to make us suffer. It is no accident that both Pearce and Lovelock include the word 'revenge' in their book titles – and the revenge is being exacted on humankind for our systematic maltreatment of the environment.

Six Degrees and Counting

Just how that revenge might manifest itself in practical terms is very graphically portrayed in Mark Lynas's *Six Degrees: Our Future on a Hotter Planet*, which ingeniously records the likely changes that would occur from each extra degree Centigrade of global temperature from 1 up to 6 in sequence. As we move up the scale the effects on our physical environment become progressively more disastrous, until we reach 6, which for Lynas marks 'the ultimate apocalypse' for humankind, where we find ourselves facing, not just widespread social and political breakdown, but 'the worst of all earthly outcomes: mass extinction' as a species.[18] To press the point home, Lynas notes that the last time life had to deal with a +6°C world, in the Permian period (over 250 million years ago), up to a staggering 95 per cent or so of all species, from land and sea both, became extinct.[19]

At +6°C humanity would appear to have little hope at all, but even at +1° (which we are fast approaching, with an increase of around 0.7°C recorded over the last century), we would be confronted by an array of problems that would tax our socio-political system very considerably. Atoll nations such as Tuvalu, Kiribati, and the Maldives would disappear, creating several hundred thousand extra refugees for the world to deal with. Tuvalu already has a treaty in place with New Zealand to take a number of the population when the time comes, although the rest are as yet unprovided for. Parts of the globe are going under now as a result of global warming, with the experience of the inhabitants of the Sundarbans delta area in the Bay of Bengal, shared between India and Bangladesh, presenting a stark warning of the future that low-lying areas of the world face. Accelerated melting of the Himalayan glaciers upstream has increased the volume of the delta's rivers, the Ganges and the Brahmaputra, to the extent that entire islands are being swallowed up in the flood season. As a local put it, '[n]ature used to give us food and crops, now all it gives us is misery, a cruel sea that covers us in sores, destroys our homes and threatens to take our families' lives. We are living in hell.'[20] The note of utter despair sounded here is no doubt one we shall become ever more accustomed to hearing in the near future.

A rise of +1°C seems all but inevitable, and most scientists concede that we are probably on course for +2°C as well (and not necessarily all that far into the future), so we have much to reflect on already. The process is no longer notional; it is under way.

Just to give us even more food for thought, there are projections that see global warming going much higher than +6°C. The World Resources Institute, for example, has forecast a +30°F (+16.66°C) increase by 2075 if no action at all, or only cursory action, is taken, which does make one wonder if the Earth could ever recover sufficiently to support a significant amount of life again.[21] We know that life has an expiry date anyway, in that the Sun will eventually burn out to become a massive red giant star unable to heat or light its planetary system, but it would seem extremely careless to have brought this event forward by several billion years. At several points in the past (the last being around 600 million years ago) there was a 'Snowball Earth' with snow and ice covering most of the planet, but a 'Fireball Earth' is far more of a threat to life in general.[22] That is within the realms of possibility if we carry on as we are doing at the moment, and +16.66°C would certainly merit the 'fireball' description. Perhaps it is not just ecocide that we are in the process of committing, but also climaticide?

The Politics of Heat

George Monbiot is similarly very worried about our prospects, regarding most Western governments as lacking the political will to take any really positive action to curb carbon emissions – in effect, accusing them of being in the pocket of big business when it comes to this issue. Neither is the general public exempt from blame as to politicians' lukewarm response to climate change. As Monbiot points out in *Heat: How to Stop the Planet Burning*, politicians have long since recognized that the public is unlikely to vote for any party promising to cut living standards drastically, as a radical approach to global warming would surely dictate; thus, '[t]he government's climate change policies often seem to fall apart when they encounter even mild opposition from either citizens or corporations.'[23] (A recent example illustrates how difficult it can be to make any move at all in this direction. In its March 2008 Budget, the British government found itself lambasted by both the parliamentary opposition and the right-wing national press for putting forward a programme of slower economic growth than of late. This was to fail the nation, opponents claimed, which took for granted that it was the government's duty to deliver increasingly high levels of growth on a permanent basis. Even a slight reduction was considered anathema to our way of life, and thus worthy of condemnation.) Despite this all too frequent turn of events, Monbiot feels there is still time to 'stop the planet burning', and he puts forward an impressively detailed argument as to how we could set about achieving this objective – while remaining keenly aware that, at base, this is a political rather than a technological problem. Unless there is a change in consciousness, then we are merely tinkering at the edges, communicating the appearance of tackling the problem while leaving it essentially untouched. Governments in particular are becoming very adept at the latter trick, as we shall be noting at various points over the course of this study; data can always be manipulated.

Much against his wishes, Monbiot too feels compelled to offer support for increased investment in nuclear power. It is a case of it being the lesser of several evils, although he cannot see it as an all-purpose, long-term solution to our energy requirements. For one thing, the topic itself polarizes debate to the extent that it can be extremely difficult to make rational assessments of the evidence: 'However much reading you do, you still don't know what or whom to believe.'[24] Nuclear power's safety has to be a worry, but whereas the Green movement is obsessed with this factor, the statistics do not always back up their more melodramatic claims as to the dangers we face from it. Monbiot also worries that the more nuclear power stations there are, then the easier it will be to use the technology to build nuclear weapons and thus pose a threat of nuclear war. Even so, he concludes that nuclear power has to be given the benefit of the doubt under the circumstances: 'the grim accountancy which must inform all the decisions we make obliges me to state that nuclear power is likely so far to have killed a much smaller number of people than climate change.'[25]

This is something less than a ringing endorsement, and Monbiot is on far happier ground when he is exploring other emission-saving options such as virtual shopping and an improved transport system based on a vastly improved intercity bus network. As events in the UK have proved, however, it is the nuclear option which has come to dominate the political agenda of late. There are few votes to be had in promoting online shopping or more luxurious bus travel; politicians prefer something altogether more dramatic and headline-gathering, and sad to say, so does the general public, it would seem.

Gabrielle Walker and Sir David King believe we can manage the heat as well, and despite their insistence that our window of opportunity to arrest the progress of global warming is probably only about twenty years or so long, they take a fairly optimistic line on our chances of survival. While admitting that some schemes, like sulphur sunshading, are best avoided, they display quite a bit of faith in technology, suggesting that low-carbon energy is within our reach. King himself was the founder of the Energy Technologies Institute, a collaboration between the UK government and the private sector dedicated to developing low-carbon technologies. Until a range of those technologies comes properly on stream the authors advocate that we invest heavily in 'carbon capture and storage' (CCS), whereby carbon from power plants is collected before it is emitted into the atmosphere and then buried underground (often in liquefied form). They claim that this method could be '[t]he most important bridging technology between using fossil fuels and new, low-carbon alternatives'.[26]

Walker and King also join in the chorus for a large-scale redevelopment of nuclear power, arguing that opposition to it from the Green movement has not taken account of the vast improvements that have been achieved in the technology since the early days. Unlike the majority of other low-carbon technologies, nuclear is ready to go now. One quibble that can be raised about such faith in technology, however, is that, as Jared Diamond has acutely observed, it makes 'the implicit assumption that, from tomorrow onwards, technology will function primarily to solve existing problems and will cease to create new problems'.[27] Technology's recent track record would suggest that is an assumption it would be very hard to defend without reservation; no one was aware of what cars and planes would do to the environment when they were developed, and nuclear power has hardly been without its problems through the years.

As for the politics, the authors are confident that Kyoto can be renegotiated, and that the World Trade Organization (WTO) could be used to ensure that all countries comply with emission targets by the threat of trade sanctions being operated against them if they do not meet their obligations. They also think the business community will be increasingly drawn into the campaign to reduce carbon emissions by the business opportunities that will arise, and that this is a trend which should be welcomed: 'The colour of money is now officially green,' they announce (although this does have the unfortunate effect of making it sound as if the planet is only worth saving if there is a profit in it for someone).[28] At a personal level we can all do our bit as well – buying the right light-bulbs and energy-efficient appliances, recycling wherever possible, cutting our car use and air travel. Taking Walker and King's side on this, even if none of these things were to achieve great savings, they would help to create a collective consciousness of the importance of acting against global warming, which could be extremely valuable in building support for more expensive measures, especially those requiring tax increases. Far from Lovelock's pessimism, this is a resolutely upbeat message: 'Above all, don't despair. The climate problem is certainly a hard one, but it's not intractable.'[29]

Nature RIP

Bill McKibben's line on climate change, however, opts for the dramatic mode. Humankind has killed off nature, which is no longer an independent entity in its own right but instead a mere offshoot of the human race's activities, particularly our obsessive desire for domination over our world:

> We are no longer able to think of ourselves as a species tossed about by larger forces – now we *are* those larger forces. Hurricanes and thunderstorms and tornadoes become not acts of God but acts of man. That was what I meant by the 'end of nature'.[30]

Climate change becomes the objective evidence of our 'success' at domination, and we have only ourselves to blame that it is now all going so horribly wrong. So engrained within us is that drive to exert control over the environment, and to enjoy the affluent lifestyle that goes along with it (at least in the West), that, in McKibben's despairing words, 'it's far too late to stop global warming. All we can do is make it less bad than it will otherwise be.'[31]

The recommendations for making it less bad are not likely to be found very palatable by most of us in the West, since they require a radical change in our mindset: a retreat from the ideals of modernity, particularly the notion that there must be continual economic growth and technological progress. We have to lose our faith that we can somehow or other engineer our way out of the situation we are caught up in, that it is just a matter of adapting, say, crops to thrive in warmer temperatures by altering their genetic structures. Rather, we have to want less in the way of material goods and be prepared voluntarily to give up a great many of the pleasures and benefits we have become used to having in a high energy-using culture: 'Possession of a certain technology imposes on us no duty to use it,' and the more we can resist the temptation, then the greater the chances of our survival in something like the form we have hitherto known.[32] As another of his books puts it, McKibben feels we should be 'living lightly on the Earth'.[33]

McKibben even flirts with the ideas of the deep ecology movement, agreeing with them that a substantial reduction in the Earth's population would be a good idea (the suggested numbers vary, but rarely exceed more than one or two billion). While he does not condone all of the policies of the radical EarthFirst! group (which can resort to violence and terrorist acts in pursuit of its aims), he is very taken by their belief that we should not consider ourselves to be above nature, that we need to develop a new sense of humility if we are to continue as a species. Unfortunately, McKibben can see little proof of this happening in society at large at the moment, and his book is a despairing cry for what we have lost in the course of imposing our will on the natural world. Nature red in tooth and claw is fast becoming little better than a folk memory, and the full implications of what it means to live in the anthropogenic age are only really starting to sink in. McKibben for one is appalled at what these are proving to be.

Data, More Data, and the Politics of Heat

The critical point to emerge from all these commentaries is that there is no clear consensus as to how to combat the problems that global warming confronts us with; that there are far more questions on offer than clear answers. As the *Stern Review* readily acknowledges, it is necessary to recognize the high degree of uncertainty involved, warning us that although '[t]he science of climate change is reliable, and the direction is clear . . . [w]e do not know precisely when and where particular impacts will occur.'[34] We have lots of data, therefore, but dwelling on them does not promote a feeling of confidence – more often than not it is one of confusion as to which set of data to believe. What do the data tell us? There are the hard data as to where we are now with reference to phenomena like average temperatures, sea levels, and carbon concentrations in the atmosphere; then there are the computer models which project the trends the data reveal into the future. I will deal with each in turn.

First, let us examine the rise in average temperatures around the globe. There is little disagreement about this; temperatures have risen significantly over the last century or so, and more sharply yet in the last couple of decades. In the Arctic, for example, one of the fastest-warming parts of the planet, the average temperature rose by 2.2°C between 1960 and 2000 alone. The data seem unchallengeable on this point, with a succession of 'hottest years in recorded history', as the press likes to herald them, arriving of late, such as in 1998. Then there was 2003, when 30,000 deaths due to heat were recorded throughout Europe over the course of the summer (15,000 of them in France alone, many among the elderly trapped in hot apartments). Forest fires are now a regular occurrence around the Mediterranean in the summer, as well as in southern California and south-east Australia, and with drought a recurrent problem their frequency and intensity is, if anything, likely to increase.

Where there is disagreement is over the cause of this temperature rise. For global warming sceptics warming is a normal and natural part of the Earth's existence (as are ice ages), and is mainly connected to its relationship with the sun rather than to human activity – an argument we shall be exploring in more detail in the next chapter. Whatever the cause, we are already beginning to find it difficult to cope with the rise, and all the obvious means of doing so merely have the effect of increasing our energy usage, and thus our carbon emissions: through the use of air conditioning, as a case in point. Mark Lynas has shown us how each added degree of temperature will affect us, so we cannot say we have not been warned.

Then there is the issue of sea levels. If these rise by even just a few metres, there will be widespread devastation around the globe, as well as to the world's economic system, which can hardly shrug off any extensive damage to economic powerhouses such as New York, London, or Shanghai. While we could in theory relocate the population from these cities to safer havens inland, it would be a massive, and phenomenally expensive, undertaking. True to form, Lomborg raises a dissenting voice, arguing that we have in fact little to fear since sea ice is already displacing its volume in the water, like ice cubes in a glass. This is true enough, but it is not the sea ice we have to worry about (as in the Arctic or Antarctic ice shelves), but the ice sheets, which are resting on the land beneath in Greenland and Antarctica. These sheets are massive in size, so that if they do melt, a dramatic rise in sea levels is inevitable – like ice falling into a glass from outside. The West Antarctic sheet, for example, is the size of Mexico and has an average thickness of 2 kilometres; that alone could raise sea levels by 5 metres if it collapsed.

Yet even the West Antarctic sheet pales into insignificance compared to the East Antarctic ice sheet, which at 10,200,000 square kilometres in bulk and 4 kilometres thick (eight times the volume of the West), contains enough water to bring sea levels up by an astonishing 50 metres. The 3-kilometre thick Greenland ice sheet would add yet another 7 metres to the total, and we really would be looking at the end of civilization as we know it if we reach that stage of global warming (+4°C in Lynas's schema). Huge areas of the globe would be under water and with them the greater part of the world's population and socio-economic infrastructure as currently distributed.

The good news is that the melting could take centuries (although, as we shall see below, that is open to dispute, as is almost every projection in this area of discourse), so there would be time to adapt to the incremental change that the process involved. Whether there would be the political will to devote the resources to make the adaptation required by constantly rising sea levels would be another matter entirely; the time scale alone would tend to militate against this, with most politicians being far more concerned about the next national election than the fate of the world's coastal population several generations into the future. The closer we get to the final collapse, the more that minds would no doubt be concentrated by it; but the closer we get, the less we shall actually be able to do to alleviate the suffering the destruction will cause. Our balancing act at the edge of chaos would be unsustainable. In the main, the politics of heat give little grounds for optimism.

Can we trust the projections about the time scale for events such as the melting of the ice sheets, however? As the glaciologist Richard Alley has pointed out, these are based on models of the process which do not really reflect the changing dynamics that global warming can bring about in poorly understood entities like ice sheets:

> We used to think that it would take 10,000 years for melting at the surface to penetrate down to the bottom of the ice sheet. But if you make a lake on the surface and a crack opens and the water goes down the crack, it doesn't take 10,000 years, it takes ten seconds. That huge lag time is completely eliminated.[35]

It is hardly just an academic point. With 10,000 years of warning, preparations can be made; with 10 seconds we are powerless to react and are plunged into an immediate crisis. The ice sheet would not melt in 10 seconds, of course, but its demise would be several thousand years closer to us than we had thought, and that is a sobering thought – how many more short cuts might nature be able to find? Lakes of the type Alley mentions have indeed been forming on the

Greenland icecap in recent years in the summer season – there is even a Greenland 'lake district' now and there is evidence to suggest that cracks are beginning to occur. Yet again we have an apocalyptic scenario to ponder, and one that could arrive far earlier than anyone had hitherto thought possible. McKibben's warning that we can no longer go on thinking of the Earth as an organism that changes 'with infinite slowness' really does need to be heeded;[36] nature is more than capable of going into 'fast forward' mode when it pleases. We can model all we want to but global warming retains the ability to keep springing surprises, not to mention bad news, on us. As the climate modeller Tim Palmer has put it, 'I don't want to undermine the IPCC, but the forecasts, especially for regional climate change, are immensely uncertain.'[37]

Other scientists are more circumspect about the impact of the Greenland lakes on the ice sheet beneath, and a study led by Sarah B. Das came to the conclusion that it might have been overestimated:

> Considered together, the new findings indicate that while surface melt plays a substantial role in ice sheet dynamics, it may not produce large instabilities leading to a sea level rise. There are still other mechanisms that are contributing to the current ice loss and likely will increase this loss as climate warms.[38]

Surface ice water was held to be responsible for only about 15 per cent of the ice sheet's movement, and the team's leader remained sceptical about its ability to find a way down to the bedrock underneath. Nevertheless, the fact that the drainage of one such lake in July 2006 – 5.6 square kilometres in size, containing 11.6 billion gallons of water – temporarily raised the ice sheet beneath and doubled its average daily rate of movement, is evidence of the power they have. That the event in question took only 90 minutes to occur does make one wonder at the stress that is being created in the underlying ice sheet, and whether the multiplication of such events, which global warming certainly promises to deliver, will generate a tipping point for the sheet's stability. While the report seems at first sight to be reassuring and to allay our worst fears, it leaves enough loose ends for doubts to start creeping back in; just how much do we understand about those 'other mechanisms' at work on the sheet, for instance? And what about the fact that the Arctic is warming noticeably faster than most of the planet is?

The discrepancy in the data in this instance is only too characteristic of the debate about climate change in general, but it has to be said that the modelling projections are rather regularly being foreshortened ('from the millennium to the decade', in McKibben's emotive phrase[39]), so we should be monitoring the progress of the Greenland lake district with some concern, the Das study notwithstanding. If the Antarctic proves to follow a similar pattern to Greenland, then we really are facing catastrophe, with unmanageable rises in sea levels the inevitable outcome. At least one eminent climate scientist is sounding a warning on the situation in Antarctica. James Hansen, the head of the Nasa Goddard Institute for Space Studies and an outspoken critic of the American government's cavalier response to climate change, has asserted that, '[i]f we follow business as usual I can't see how west Antarctica could survive a century. We are talking about a sea level rise of at least a couple of metres this century.'[40] If that sounds bad news, Hansen's follow-up prediction is that if all the ice on the planet melts, and he thinks this is a real possibility the way carbon emission levels are remorselessly rising, then the sea level rise would be 75 metres, 'a guaranteed disaster', as he bluntly puts it.[41] It would be hard to disagree with this particular contribution to the 'apocalyptics' genre. The physical, social, and political landscape would be changed beyond all recognition by such a development, and Lovelock's vision of remnants of humanity eking out a miserable existence on remote parts of the planet would most likely have become the harsh reality.

Even if the more optimistic projections prove to be correct and the melting of the ice sheets does take centuries, sea levels will still rise in the mean time by virtue of thermal expansion. Water expands as it becomes warmer, and it has been estimated that the kinds of rise we seemingly cannot escape – around +2°C, for example – would on their own raise the sea levels globally by 1 or 2 feet.

Bizarrely enough, however, we might start to hear some arguments in favour of Arctic melting from interested parties. Recent studies by geologists have revealed substantial oil deposits beneath the Arctic Ocean, but these cannot be exploited properly until the ice melts, which would be evidence of advanced global warming. At which point, new supplies of one of the major causes of global warming could come on stream, accelerating the process massively. We have to hope that even the oil companies can see the madness of getting ourselves into that position.

Carbon concentrations in the atmosphere have increased from their pre-industrial level of 280 parts per million (ppm) to 380 ppm. The point at which rapid temperature rise is triggered is still a matter of some dispute, but it is generally thought that anything above 450 ppm will be highly dangerous, and that we should do all we can to prevent that level being reached. After that point, acceleration sets in, with positive feedback becoming a progressively more important factor. Unfortunately, we are heading towards that target at a rate of 20 ppm per decade – and the rate is steadily increasing. Kyoto was meant to start the process of reduction but, as we have seen, it has had precious little effect so far, and is unlikely to do so in future either, unless the USA joins in wholeheartedly. Even then, we have the problem of whether countries are really meeting the targets that have been set for them or are massaging the figures to their own advantage. As I will be discussing in Chapter 12, the UK government has been found guilty of the latter sin of late. Without a more rigorous regime of both monitoring and enforcement it is hard to see any real progress being made in this area; but equally, it is imperative for all the world's countries that it is.

Another major contributor to climate change is likely to be the world's peat bogs and permafrost regions. These produce methane as they thaw, and methane is a gas far more dangerous than carbon dioxide in that it is a hundred times more powerful in terms of its warming qualities (cattle contribute to the methane emission total too, so humanity's seemingly insatiable appetite for hamburgers is also a problem of note). The world's considerable acreage of permafrost has been largely stable in our era, but climate change is beginning to melt it in areas like Siberia at a disturbing rate (and just to get a sense of the scale involved, the West Siberian peat bog alone is as big as France and Germany combined in size). If this continues, then methane levels in the atmosphere will rise alarmingly, triggering much faster climate change in turn. In periods in the past when the atmosphere was rich in methane, the effects on the environment were devastating; 55 million years ago a massive release of methane from the ocean – over a trillion tones, it has been estimated[42] – raised global temperatures by as much as 10°C, destroying two-thirds of the ocean's species in its aftermath.

The destruction of huge peat swamps in places like Borneo, in order to replace them with more profitable farmland for cash crops (often to be used in the production of biofuel), releases vast amounts of carbon rather than methane, and has already contributed to a major environmental crisis throughout south-east Asia. In 1997 a huge cloud of smog hung over the entire region for months, largely thanks to peat bog clearance operations. The lure of profit is yet again putting us all at risk; as Will Hutton points out, markets are notoriously 'myopic' and invariably 'overvalue the immediate'.[43] What that all too often means nowadays is an immediate contribution to the store of greenhouse gases in the atmosphere.

Planet Death Versus Electoral Death

It becomes clear from considering the work of theorists such as the above that the commitment to economic growth is the major barrier to tackling climate change. Economic growth remorselessly drives up the carbon levels in the biosphere, and no nation seems willing to alter its policies significantly in this respect, a few sops to the Green movement notwithstanding. It is now taken to be all but a human right for living standards to improve steadily on a permanent basis. Politicians consider any call for an economic slowdown to be tantamount to electoral death – and as the reaction to the UK government's March 2008 budget above suggests, they are most probably right. That means we cannot take refuge in simply blaming the professional political class for our plight; we are all implicated, particularly those of us in the West where such ideas are most deeply engrained in the public consciousness (although China is beginning to catch us up on this score). We have to start asking ourselves why it is that we cannot countenance even a relative slow-down in economic progress, why we are so resistant to adjusting our expectations to fit a rapidly changing world situation.

From this we might conclude that it is the free market that is the real culprit, and the way that it currently operates really does need to be overhauled radically. This is an issue we shall be addressing in more detail in Chapters 4 and 12; suffice it to say for the time being that if climate change incontrovertibly goes past a key tipping point (truly damaging rises in sea level on a systematic basis, say), and both politicians and the general public come to recognize that planet death trumps electoral death, then we will have to be exceedingly careful about what action is taken to deal with this worst-case scenario. First, however, let us consider what arguments there are to counter the case being made for global warming: what the 'coolers' have to say against the claims of the 'warmers', and how much credibility they can be accorded.

3
Global Warming: The Arguments Against

So much for the time being for the arguments for global warming; what do the sceptics say? Various arguments have been outlined by climate change sceptics: such as, that warming is part of the Earth's natural cycle (and they have some persuasive data from the past to back up this claim); that it is due to solar activity (sunspots or the solar wind, for example); that the data on which the climate change case is predicated is flawed, or at least open to different interpretation from that it is receiving from climate change supporters (always a problem with modelling, which can be made to yield pretty much what the modeller wants); that the problem is being overstated and that methods such as offsetting will be enough to solve it. Failing all that, it has been argued that technology will find an answer, as it so often has ridden to our rescue in the past: although on that point, one scholar of the rise and fall of civilizations has somewhat acidly remarked that it is little better than 'a "faith-based" approach to the future'.[1] It has even been asserted by some sceptics that global warming is an elaborate 'scam' by the scientific community designed to win research grants for pet projects, and that the public should be made aware of how they have been duped (at least about the probable impact of any warming there might be on our way of life, anyway). The scepticism is generally not so much about the fact of warming, therefore, but the causes and then the likely future effects, with what Jared Diamond has called the 'non-environmentalists' arguing that the effects will probably be beneficial overall.[2] It does have to be noted, however, that some arguments are still being pursued that the planet is actually about to cool down rather than heat up, and I will consider the evidence for these too in due course.

The issue does arise fairly immediately as to whether such arguments are compromised by their funding sources – which are very often the international oil companies. Many foundations and research institutes that issue sceptical reports on global warming and its effects turn out to be funded by large organizations such as ExxonMobil, and this can put a completely different spin on the role of those sources in resisting the climate change case; special interests manifestly have to be taken into account in this debate. Another issue that arises in this context is whether the arguments of the sceptics provide an excuse to go on exploiting the environment recklessly (clearing Amazonian rain forest on a massive scale, for example), thus exacerbating the problem further. Even if humanity were not the main cause of global warming, and it really was to be traced back almost exclusively to the sun and its cycles, it would hardly make sense to do things that added to the process. Why throw oil on a raging fire?

We need to compare the work of oil company-funded climate change sceptics and environmentalist sceptics such as Bjorn Lomborg. Although the latter is not strictly speaking a denier of climate change, he feels its predicted effects are grossly exaggerated and that reducing carbon emissions will be both prohibitively expensive and largely unnecessary: thus his styling of himself as 'the skeptical environmentalist'.[3] He is vehemently opposed to the Kyoto Protocols, and in his books he campaigns instead for a raft of practical measures that, if put into practice, would improve the quality of life of the world's population overall – reducing poverty and fighting diseases such as AIDS and malaria with existing drugs, for example, rather than investing heavily in prohibitively expensive technology of often unproven effectiveness. Lomborg's ideas merit careful consideration, although his tendency to see the global warming case as something of a conspiracy on the part of self-interested politicians and scientists is highly questionable; it is not difficult to see why the denier lobby has appropriated him to their cause, for all his undoubted environmental sensitivity.

Temperature Change in History

There is no dispute that the Earth has been subject to wild swings of temperature over its history, and that a range of internal and external factors, all of them still capable of being activated by events, can be identified as contributing causes. Internally, the Earth is still a very unstable system, and earthquakes and volcanic eruptions can have effects destructive enough to affect climate very markedly. These often can be long-term in nature; the eruption of Mount Tambora in Indonesia in 1815, for example, triggered what came to be known as the 'year without a summer' in 1816. Volcanic dust in the atmosphere from this eruption – plus several others that occurred around the same time in what was a period of unusually intense volcanic activity – substantially reduced incoming solar radiation, thus shortening the growing season worldwide and leading to a complete failure of the harvest in areas like southern Germany. Widespread social unrest duly followed throughout Europe. The supervolcano underneath Yellowstone National Park in the USA is being looked at of late with a certain amount of trepidation by earth scientists, well aware of the devastation this could unleash, not just over America or the northern hemisphere, but potentially the entire globe, if it were to erupt. Surveys of its history tend to suggest that it is overdue for a major eruption, so that is something else for the 'panicologists' amongst us to fret over. Externally, there is the sun to be held responsible, and we will come on to consideration of its cycles later.

Even in recent times wide variations in temperature can be noted, with the Medieval Warm Period (c. 800–1300) and the Little Ice Age that followed (c. 1300–1850) being well documented in the historical records.[4] (It may come as something of a surprise to discover that we are technically still within 'the Great Ice Age', going back over a million years, currently undergoing an interglacial – although humanity seems to be doing its best to alter the structure of the cycle.) Within a few hundred years we move from vineyards flourishing in England – a 'climatic golden age' as Brian Fagan has described it, although probably somewhat cooler overall than today[5] – to the Thames freezing over to the extent that it was solid enough each winter for fairs to be held on its surface. (The continental experience of this latter event can be gleaned from the popularity of paintings of ice scenes in the work of the Dutch school of the period – painters like Hendrik Avercamp (1585–1634), for example, where we find the ice crowded with inhabitants of the local town, all engaged in some form of activity, whether leisure or commercial, on the frozen river beneath the town walls, the seventeenth century being the coldest period of the Little Ice Age.) Neither of these events is held to be a result of human activity; apart from the fact that the medieval period was a pre-industrial age, the human population of the globe was radically smaller than it is now (it is estimated that even as late as 1500 it was only around 460 million as opposed to the current 6.7 billion). Humanly generated carbon emissions were hardly a problem under those circumstances, and indeed a recent study of the Little Ice Age rather worryingly suggests that rising temperatures are as likely to push up CO_2 levels as the other way round (as has been the received wisdom).[6] Yet again we are forced to acknowledge how tricky it can be to establish clear cause-and-effect patterns when dealing with the environment; uncertainty seems to remain the normative state.

In human terms the Little Ice Age was a substantial event, covering a large chunk of pre-modern history and even some of the modern era – the Thames was still freezing over in Victorian times, making the climate differences that are now being experienced in the UK in the early twenty-first century all the more remarkable to observe. Apart from anything else, vineyards are a commercial proposition once again, at least in the south of the country, for the first time since the fifteenth century (wine writers even speculate on them going as far north in Europe as Finland if current trends continue unabated, which would beat the Medieval Warm Period's furthest outpost of southern Norway). Unlike the Medieval Warm Period, however, we now have an industrialized civilization, and 6.7 billion people globally to consume its products.

Less helpfully for the climate deniers' case, as well, is the fact that entire civilizations did actually collapse at least in part as a result of such events as the Medieval Warm Period: the Tang dynasty in China, for example, and the Maya in Central America. As authors such as Brian Fagan and Jared Diamond have made clear, we are only the latest in a line of sophisticated cultures to find themselves facing the prospect of internal collapse from the extra stresses imposed by large-scale global climate change.[7] It cannot be assumed that we are immune to the fate that befell societies like the Tang or the Maya when they found themselves confronted with unprecedented conditions, such as extended drought, that outstripped the technology at their disposal and thus their ability to adapt. Sometimes, depressing as it may be to admit it, there just are no solutions and you have no alternative but to accept your fate. Fagan has remarked that, when looked at from a global perspective, 'it is tempting to rename the Medieval Warm Period the Medieval Drought Period.'[8] We seem well on the way to similar conditions nowadays in places like the Sahel, Australia, and the American south-west, where drought has become the normal state of affairs – many parts of these see no rain at all for several years at a stretch. As Joseph A. Tainter has remarked in his pioneering work on the topic, '[h]owever much we like to think of ourselves as something special in world history, in fact industrial societies are subject to the same principles that caused earlier societies to collapse'[9] – and climate change looms large in the process.

The sun and its various cycles of activity are the major factors to be taken into account when looking at pre-Anthropocene era climate change. It is well known that the sun goes through changes that affect the Earth's climate, with sunspots being their visible sign. Sunspots nowadays are observed to work on an eleven-year cycle over which they increase and decrease in number, thus altering the amount of solar radiation that reaches the Earth. But sunspot activity declined markedly during the Little Ice Age, and in fact seems to have all but disappeared in the seventeenth and early eighteenth centuries. The Victorian astronomer E. W. Maunder noted from the records that from 1645 to 1715 sunspots were very rare, and from 1672 to 1704 apparently non-existent. The 'Maunder Minimum', as this has come to be called, coincides with some of the coldest spells experienced during the Little Ice Age. Although the exact link remains elusive, there are, as Fagan suggests, 'compelling connections between the prolonged periods of low solar activity and the maxima of the Little Ice Age'.[10] Solar activity has on average been rising since the Maunder Minimum, but we have no way of knowing whether such anomalous periods might occur again without warning. Neither are we clear about what causes variations in the solar wind, the stream of particles emitted by the sun which are also known to affect climate on Earth (in addition to being the cause of the aurora borealis, which was also very rarely sighted during the Maunder Minimum).

It would no doubt be to our advantage to experience something like another Maunder Minimum to counter global warming, although whether it would have as much of an impact on a planet as deep as ours is in the throes of the greenhouse effect as it did in the Little Ice Age is a moot point. Given how fast the world has warmed up since the end of the Little Ice Age in the mid-nineteenth century, it would hardly be a long-term solution anyway, even if it might boost the case of the coolers for the time being. Human beings cannot be held responsible for what happens on the sun, so we can put both the Medieval Warm Period and the Little Ice Age down to natural causes. Yet we might also care to reflect on the fact that, although sunspot activity has actually been getting less since 1980, the planet is still warming at an increasing rate. So the evidence does not tend to suggest that we can rely on the sun to bail us out.

To Dimly Go

An intriguing aspect of our pollution of the atmosphere is that it has led to global dimming, which on the face of it should be keeping the Earth's temperature down. After studying records of sunlight from around the globe, the scientist Gerry Stanhill discovered a decline of the amount received of 1–2 per cent globally for each decade from the 1950s to the 1990s. The problem was not the carbon dioxide being released by fossil fuels but the other particles that went along with it, such as soot and ash. These particles had been altering the composition of clouds so that they were reflecting more sunlight back into space, thus reducing the world's rainfall and triggering droughts. In other words, the particles were cancelling out some of the effects of the carbon dioxide that was being pumped into the atmosphere in ever-increasing quantities by our use of fossil fuels. The bad news is that the pollution from the particles is systematically being reduced by improvements in technology, setting up the possibility of a surge on the warming side as the cooling effect declines: 'we'll get reduced cooling and increased heating at the same time and that's a problem for us,' as the climate scientist Peter Cox has warned in what can only be described as an understatement.[11] The relatively modest 0.7°C increase we have already experienced could start soaring quite soon, in what could be considered one of the more ironic developments in the campaign to bring our carbon footprint down. Who would have thought pollution would come to be seen as our friend? Or that eradicating it could ever turn out to be against our best interests?

Tipping the Planet

There are also periodic variations in the Earth's axis which can affect climate. The planet's axis of rotation is not perpendicular but tilted, and it shifts between a maximum tilt of 24° and a minimum of 22.5°. At present it is at 23.5° and decreasing gradually towards its minimum on a 40,000-year cycle. There has been speculation that the tilt was caused by a collision with another planet around 5 billion years ago. Some scientists have postulated a link between these variations and ice ages, with the glaciation occurring when the tilt is in its minimum phase, and the warm periods when at its maximum. Peter Huybers of the Woods Hole Oceanographic Institution in Massachusetts explains: 'The apparent reason for this is that the annual average sunlight in the higher latitudes is greater when the tilt is at its maximum,' thus initiating the melting of the ice sheets in the polar regions.[12] This could be construed as good news by the coolers, particularly since we are perceived to be on the downward cycle to a minimum, with a fair distance to go yet. Another ice age would seem to beckon.

The Earth's orbit around the sun has also been suggested as a possible source of the process of glaciation, since this too undergoes variations, this time on a cycle of 100,000 years. Or the glaciation might be the product of some as yet unknown interaction between these two cycles. Huybers and his co-researcher Carl Wunsch have even speculated that the Earth is gradually cooling, because their studies indicate that the melting has not been as widespread as would have been expected in recent maxima. That could be even better news for the coolers, although it has to be noted that global warming is happening at a much faster rate than either of these cycles, and if we are only a third of the way through the smaller cycle that sounds like an unequal contest. Global warming is a matter of centuries, or even just a few generations, rather than thousands of years. Still, it undeniably gives more weight to the coolers' argument, and the warmers would have to concede that there is a reasonable basis for such speculation.

A related point that the coolers could use to undermine the more apocalyptic scenarios about sea level rises, is that some studies suggest that the melting of the Arctic and Antarctic icecaps could alter the planet's centre of gravity, thus actually reducing the sea level at various points around the globe (Cape Horn and Iceland, for example).[13] Bill McKibben reports himself 'awed by the idea' of the Earth 'tipping' in this way, but what its overall effect would be we could hardly say;[14] even if it does show how modelling can vary in its predictions, it seems a dangerous gamble to make that this particular one would win out over the many others that forecast unmanageable rises instead. But again, it would have to be granted that there is at least a basis for speculation of this kind.

The Sea to the Rescue?

The coolers might take heart as well from a recent study claiming that, far from there being a prospect of temperature rises, '[t]here could be some cooling' instead in Europe and North America for around a decade, thanks to the fact that the meridional overturning current in the mid-Atlantic, which carries warm water from the tropics to northern latitudes, appears to be slowing down.[15]

While this may delay some of the effects of warming, it would not be enough to arrest it overall, and we cannot rely on a series of other such unexpected phenomena to save us. It might also lead to a sense of complacency that would not be helpful in the longer term. What if events elsewhere were to counteract the change in the meridional overturning current? What if the current were to speed up again? If it did we would be in danger of the sea losing much of its oxygen, as happens when warming occurs, and the less oxygen there is in the water, then the less life; the oceans would be turned into 'oxygen deserts'.[16] Even so, the study on the current is further proof, which the coolers will always be happy to exploit, that nature is not as predictable as we would like to think, and that we still do not have anything like comprehensive knowledge of the way the Earth system works.

Whether the effect of the meridional current will be offset by changes elsewhere in the system is always a point to bear in mind as well. Just after the details of the current study were released, another report from the US National Oceanic and Atmospheric Administration revealed that carbon dioxide levels, as measured at the Mauna Loa Observatory in Hawaii (one of the best sites in the world for conducting these tests), had just hit their highest level for 650,000 years and were increasing far faster than had been expected.[17] This led some scientists to conjecture that the Earth might be losing its ability to absorb CO_2 as efficiently as in the past, the suggestion being that the planet's 'natural sinks', such as the forests and oceans, were struggling to mop up the ever-increasing amount of greenhouse gases in the atmosphere. The unpredictability of the Earth system can just as easily undermine as bolster the coolers' case, in what can soon turn into an elaborate game of claim and counter-claim.

The Carbon Offsetting Industry

Carbon offsetting has been championed for some time as a relatively painless way of reducing emissions and reaching the highly desirable 'carbon neutral' state, where emissions are effectively being cancelled out by carbon-absorbing actions taking place elsewhere.

Cities and countries vie with each other to achieve this goal, and a sizeable industry has grown up to expedite offsetting; the market can always spot a new niche and 'carbon capitalism' is doing its best to exploit it. Planting trees has been one of the major features, and over the years this has developed into a lucrative business (the carbon offsetting market in general was estimated to be worth around $60 billion in 2007), particularly in the developing world, where there is greater access to unused land than in the more intensively organized and cultivated West.

Lately, however, some very searching questions have been asked about the impact of carbon offsetting and whether its promises can be upheld. Critics have claimed that it is a delusion to think we can offset carbon as easily as proponents claim, and that activities such as tree planting are a mere sop to our consciences. The effectiveness of trees in locking up carbon is now a matter of dispute, with some scientists suggesting that there is a saturation level past which trees cannot go in absorption. There is now strong evidence to suggest that one of the effects of climate change is to slow down tree growth, which, as the science writer Douglas Fox has pointed out, means that trees 'might not be able to lock away our CO_2 after all'[18] (the US National Oceanic and Atmospheric Administration report mentioned above would appear to provide further corroboration of that conclusion). As Fox goes on to insist, '[n]one of this is to say that we should not preserve tropical forests . . . or not replant them where they've been cut down,' just that 'we should not expect those trees to perform miracles.'[19] Such findings ought to throw into doubt the future of carbon offsetting as a business, given that it was based on the premise that trees could indeed perform such miracles on our behalf in cancelling out our personal carbon emissions. Bit by bit the easy methods of reducing carbon emissions are proving to be illusory, and we are having to face up to the fact that there is after all no painless route to carbon neutrality. The warmers would claim that neutrality was not enough anyway, that there has to be a reduction in real terms.

Carbon credits, whereby nations with low emission rates can sell allocated emission units to nations with high emission rates, is another method that has been touted as a means of allowing the West to continue much as usual in terms of energy usage. On the face of it this seems to be an idea with much to commend it, particularly if the process is priced in such a way that the developing world really does gain from it. This makes the assumption, however, that a quota system alone will resolve our difficulties, whereas critics argue that everyone has to bring their emissions down, and continue to do so into the indefinite future, if we are to avoid disaster. Buying unused emissions hardly seems in the spirit of changing our entire attitude to energy usage, more like a case of rearranging the deckchairs on the Titanic; the emissions would be better remaining unused altogether. It is just one more instance of the West searching for a way in which it can resist addressing the problem with the necessary vigour and seriousness, and we really must move past that mind-set if we are to make any headway on the problem in general.

One does suspect, however, that carbon trading is an idea which will continue to exert an appeal, since it does not ask for anyone to take any concrete action – nothing has to be rebuilt, or projected into the atmosphere to unknown effect. Politicians will always find that congenial, as well as the fact that it can also be turned to account in claiming the moral high ground; the rich countries could argue that they are improving the economies of the poorer, at no cost to the latter in terms of production. It would depend on what the agreed rate for the credits was, but the developing world would seem to hold a good hand in this particular deal. We may well come to regret the lure of the easy option, however, as it simply delays the day when we can no longer ignore the size of our carbon footprint and the effect it is having on the environment – and delay merely makes dealing with it that much harder.

There are more positive examples of offsetting to report, however, as in the activities of the non-profit organization Climate Care, which is currently engaged in a large-scale programme of replacing coal-burning stoves in rural China with stoves based on fuel from renewable sources such as biomass:[20] an example of the Clean Development Mechanism (CDM) introduced in the Kyoto Protocols. (We might just note, however, that the latter measure will no doubt help towards reducing global dimming, which we now realize is at best a mixed blessing.) More reliable methods of offsetting such as this will presumably be developed in future as climate scientists begin to work more closely with their peers in other fields, both inside and outside the scientific community. As Fred Pearce has pointed out, '[e]conomists and energy technologists; geologists and architects; electrical engineers and flood managers; crop scientists and disease epidemiologists; company lawyers and even City traders – they all now find climate change is central to their work.'[21]

This kind of cooperation will become increasingly necessary in order to prevent the market alone from setting the agenda, often using very dubious arguments to attract customers, arguments relying far more on emotion than scientific rigour. Offsetting will have to do more than just salve the individual or corporate conscience by what has been called 'the sale of promises';[22] it has to be demonstrably in the public interest and scientifically provable to be so – after all, sometimes promises are not kept, and sometimes they are just plain false. The further that offsetting is removed from the market the better, in fact, and non-profit organizations are deserving of our support in this regard. When profit enters the scene it is rarely to the benefit of the environment; as a former director of the environmental pressure group Friends of the Earth has put it, to think that carbon capitalism will be an aid in the struggle against global warming 'is to believe in magic'.[23] Nicholas Stern, like Joseph Stiglitz a one-time chief economist at the World Bank, has been even more withering in his criticism, describing climate change as 'the greatest market failure the world has ever seen'.[24] (One might note in passing that, although the World Bank may be famed for its ideological rigidity in recent years, it nevertheless does seem to have the capacity to inspire a considerable amount of dissent in its upper level of management.)

A Few Cool Thoughts

How about the arguments, favoured by some sceptics, that the Earth is cooling down rather than warming up? How credible are these?

Once again, the sun's cycles hold the key, and the data do indeed suggest we are entering a cooling phase that could last for some time. Not surprisingly, climate change sceptics have been only too happy to broadcast this information widely, arguing that it throws the modelling projections of the warmers into disarray. As far as the sceptics are concerned, it shows there is no need to take action against global warming, and that we can carry on using fossil fuels and developing an even higher energy-dependent society; the sun will save us. But all that it really means is that temperature rises might prove to be slower than predicted, and even then there is no guarantee, given nature's well-attested 'fast forward' skills – skills that could be triggered by a reversal of global dimming, as a relevant case in point.

When the cooling cycle does end, therefore, we shall be even worse placed to prevent disaster if we have not succeeded in reducing carbon emissions in the interim. What we ought to be doing instead is taking advantage of this natural window to prepare ourselves for the adverse conditions to come when the sun will start rapidly accelerating the effects of humanly induced, continually increasing carbon emissions; but for climate change sceptics that would be just so much wasted effort – and resource.

Lomborg: The Skeptical Environmentalist

For Bjorn Lomborg, however, there is neither a resource nor an environmental crisis to worry about. In *The Skeptical Environmentalist* he argues that careful study of the relevant statistics reveals that many of the claims of the environmental lobby are questionable, often founded on prejudice rather than facts. Statistics become for Lomborg a means of challenging that group, and the media who support them, so that there is 'a careful democratic check on the environmental debate, by knowing the real state of the world – having knowledge of the most important facts and connections in the essential areas of our world'.[25] Those facts and connections reveal to Lomborg that things are actually getting better, '[m]ankind's lot has actually improved in terms of practically every measurable indicator', but that this is not 'good enough' if significant percentages of the world's population are still suffering badly from poverty and even starvation.[26] He insists that we have both the capability and the will to improve this state of affairs, but that we could be doing so more quickly and efficiently.

Overall, Lomborg offers a cautiously optimistic assessment of the state of the world that is concerned to avoid the apocalyptic tone of so many commentators (and he is in sharp contrast to the ones we considered in Chapter 2). In his view there are more global success stories to report than are generally credited in the media. There may still be problems, but we can work our way out of them. Some of us may be a bit queasy about his examples of success: 'if Burundi with 6.5 million people eats much worse whereas Nigeria with 108 million eats much better, it really means 17 Nigerians eating better versus 1 Burundi eating worse – that all in all mankind is better fed'.[27] Yet Lomborg himself is aware there can be an inhuman quality to such a use of statistics. His point is that it depends which examples you pick, and he accuses the pessimists of ignoring the success stories in favour of emotive cases – starving Burundians, for example – that reinforce their world view. Lomborg's belief is that if we look hard at the statistics behind global trends, we will find that they generally undermine the position of the pessimists. Essentially, his argument is that the pessimists are over-generalizing from very specific cases, and that the statistical evidence shows that we are not really facing any full-scale crisis in either our supply of resources or the state of the environment. To claim ecocide is to indulge in scaremongering.

The same approach is carried over into *Cool It*, which concentrates more specifically on the phenomenon of global warming. Lomborg's argument is based on the premise that, although global warming is a fact of life, it is neither as big a problem as so many are claiming, nor the most important issue facing humanity today:

> Statements about the strong, ominous and immediate consequences of global warming are often wildly exaggerated, and this is unlikely to make for good policy . . . I will argue that we are wrong in making climate change our primary focus. We need to get our perspective back.

> There are many other and more pressing problems in the world where we can do much more good, for people who need it much more, ultimately with a much higher chance of success.[28]

Lomborg contends that most solutions for reducing CO_2 would be both prohibitively expensive and of little long-term impact, and therefore that we should reconsider how we approach climate change overall: 'we must stop thinking about quick and expensive solutions but rather focus on low-cost, long-term research and development.'[29] He has a point about the cost, if not necessarily the impact; but I will deal with his arguments against the warmers in this section, then turn to his suggested candidates for research and development in Chapter 6, 'the smarter solutions' as he dubs them, when I will be investigating technological solutions to climate change in more detail.[30]

Lomborg is particularly critical of those peddling worst-case scenarios, as in the case of the American ex-Vice President Al Gore, whose film about global warming, *An Inconvenient Truth* (2006), has been shown worldwide in recent years, generally gaining plaudits from the scientific community (plus an Academy Award in 2007 for Best Documentary).[31] Lomborg is dismissive of such scenarios, which he considers to be little better than doom-mongering for its own sake. When their claims are investigated closely, they prove to be far less conclusive than they seem at first sight. Taking an emotive example, the effect on Arctic polar bears of global warming (and by now we have all seen photographs in the press of stranded polar bears on ice floes), he points out that in most cases the population is actually growing, and that hunting is a far greater threat to the bears than climate change. If we stop hunting bears, therefore, we shall achieve more to stabilize the population than if we spend an enormous amount of money following the Kyoto Protocols in the hope that we can arrest climate change in the Arctic and protect their natural habitat. The latter is what he calls a 'feel good' strategy, and the former a 'do good'.[32] Worst-case scenarios like Al Gore's promote 'feel good' policies; Lomborg wants us to resist that impulse, which he argues produces schemes that are invariably both inefficient and impossibly expensive, and to opt for 'do good' instead.

Gore is particularly taken to task on the subject of sea levels, with his attention-catching claim that these will rise by 20 feet or so, thus inundating cities such as Miami, Beijing, and Shanghai, not to mention the whole of the Netherlands. In reply, Lomborg cites one study by the IPCC that suggests that the Antarctic ice sheets may actually increase in size, thanks to the increased precipitation that global warming could cause, thus actually reducing sea levels. Gore's major source of evidence for his film is the work of Jim Hansen, who is also accused of overstating the case, with Lomborg, reasonably enough, pointing out how Hansen is deliberately refusing to accept IPCC projections regarding likely sea level rise over the next century. In Hansen's own words, 'I hope those authors are right. But I doubt it.'[33] Hansen is an eminent scientist who has done much to bring global warming to the attention of not just the general public but also of a sceptical American government that seems constitutionally incapable of doing anything to harm the country's economic interests, so we should be prepared to listen to his doubts; but at the very least they allow some room for scepticism in return, which Lomborg is only too happy to provide. Again, it becomes a question of how you interpret the data, and whose interpretation you think most plausible.

The IPCC comes in for criticism for good measure, with Lomborg arguing that it has become increasingly politicized of late, leading it to make more dramatic claims about humanity's role in climate change than hitherto. Initially cautious about causes, the IPCC now holds humanity largely responsible. Lomborg finds this unacceptable:

> When scientists – without new science [as the IPCC had conceded] – 'sex up' their message, it is no longer just science. It is advancing a particular agenda, namely that their area is more important for funding, attention and rectification than it really is. Sending a stronger message to politicians is simply using science to play politics.[34]

This is a fairly common objection from the sceptical side: that global warming is a nice little money-spinner for the scientific community, and that they will say whatever is necessary to keep it that way. The more that a political agenda intrudes in this manner, Lomborg contends, the less debate there is about the validity of the data we are presented with – and he believes there is a considerable amount of doubt about that validity. The longer we allow ourselves to stay in thrall to 'some scientists making scary scenarios' (a nod to conspiracy theory here, perhaps?), the less progress we shall make on the 'social priorities' that Lomborg thinks really counts.[35]

Global Unreason

Nigel Lawson, the ex-Chancellor of the Exchequer in Margaret Thatcher's government in the 1980s, is another to question whether the situation really is as bad as the warmers are insistently claiming. For Lawson, the IPCC is a 'global quasi-monopoly of official scientific . . . advice' which is not to be trusted, and the *Stern Review* on the costs of climate change is 'essentially a propaganda exercise' on behalf of a government which has already made up its mind on the topic.[36] Given that many commentators have criticized the IPCC as overly conservative in its estimates – they 'tend to stay behind the science rather than get ahead of it', as the authors of a recent book have put it[37] – this is a bold challenge to make. Global warming is dismissed as little better than a form of collective hysteria, with Lawson warning that, '[w]e appear to have entered a new age of unreason, which threatens to be as economically harmful as it is profoundly disquieting.'[38] Like Lomborg, Lawson feels that the projected costs of dealing with global warming are so exorbitant that we would be better advised to concentrate on smaller-scale interventions, such as conservation of our water resources (although he does express some doubts as to the many scare stories about scarcity there). Essentially, he recommends a 'wait and see' approach to the problem, arguing that we should respond to climate change only when events require us to, rather than spending vast amounts of money as a preventative measure. A reviewer of the book found a neat rejoinder to this attitude, however, when he noted that '[m]ost people don't expect their house to burn down. But they take out fire insurance, provided it is available at a sensible price, to protect themselves against the possibility.'[39] The analogy seems only too apt; we should be prepared to take out fire insurance for the planet.

Unreason is an emotive term; what evidence can Lawson produce to prove that so many eminent minds in the scientific and political world have fallen under its sway? In his view the IPCC has 'mutated . . . into something more like a politically correct alarmist pressure group' with which governments are afraid to be seen to be in dispute.[40] Like Lomborg, he makes the point that the panel's public pronouncements about the severity of the situation go much further than the data in their reports would warrant, thus creating a situation where debate is stifled (the author's foreword makes a point of telling us that a string of British publishers rejected the book because of its perceived unorthodoxy on the subject). He also speculates that there may be a subversive political agenda at work here, with environmentalism becoming the new home for left-wingers cast adrift by the collapse of Marxism and socialism as internationally meaningful political movements: 'For many of them, green is the new red.'[41] It is in this group's interest to blame the market economy for climate change and to paint as depressing a picture of the future as possible. Their ultimate goal is a resurrection of the authoritarian state.

Lawson remains a strong defender of the free market, and it is no surprise that he is so critical of radical environmentalist views and of Stern, for whom global warming is very largely the product of market forces being given their head (although Stern feels that these same forces can also be enlisted in the cause of green technology[42]). It will be through the free market that we adapt to climate change, with Lawson recommending the imposition of a carbon tax to discover just what both the public and the private sector are willing to pay to reduce emissions. In effect, the market will decide – but not all of us will have the faith in the wisdom of its workings that a disciple of Milton Friedman has.

Funding Scepticism

As mentioned above, the funding of the sceptical lobby does need to be taken into account when weighing up their arguments and recommendations. One of the leading voices in this field is the climate scientist Richard Lindzen, and he speaks of there being a 'climate of fear' in which sceptics are having to work.[43] But that has not stopped the major sceptics being able to continue their work. Oil companies have been generous in their funding of organizations and institutes sceptical of the human effect on climate change, and they make extensive use of their findings in defending their own activities. A notable sceptic to have benefited from oil company funding is the palaeogeologist and oil exploration consultant, Martin Keeley, who has gone on record as claiming that 'global warming is a scam, perpetrated by scientists with vested interests.'[44] Granted, there are many large-scale projects under way at present that are studying global warming and churning out vast quantities of data about the process, but the likelihood of almost the entire scientific community being party to a scam, or even just being taken in by it, is not very high. It seems churlish to question the sincerity of the scientists working on this problem; for a start, they constantly seem to be calling each others' modelling methods into question, and there is wide variation in the interpretation of the modelling exercises.

Being in the pay of an energy multinational does not mean that your scepticism is not sincere either, of course, and there is room for reasonable doubt as regards the data on global warming and its interpretation. The warmers are capable of special pleading for their pet theories, as well as imaginative leaps (neither of which is the same thing as perpetrating a scam, however). Perhaps too many scientists are hooked on apocalyptics and go further than they should in their rhetoric. As long as this is what the sceptics are telling us, then they deserve to be taken seriously. Clearly, this is all to the oil companies' advantage, and it is what they expect their funding to generate. The extent to which this predetermines conclusions certainly needs to be kept under close review.

Learning From Climate Change Scepticism

Global warming sceptics have some good arguments, therefore, and some dubious ones, although their funding sources must cast a certain shadow over their claims. They cannot always be regarded as entirely independent thinkers if the fossil fuel industry is backing their research projects, and these are not always scientifically sound either; lack of peer-reviewed status has to raise a considerable degree of doubt over their findings, and that is not an uncommon occurrence in this area of enquiry (Pearce cites some of the most prominent of these in *The Last Generation*[45]). Such qualifications notwithstanding, sceptics are nevertheless useful, in an agonistic sense, as long as they keep the rest of us scrutinizing our own position closely – and permanently aware of what is at stake. The warmers are heavily reliant on modelling, and as we have seen, this has its limitations, never mind its contradictions, and those problematical aspects of the process need to be drawn to our attention on a regular basis. Science can provide us with the data, but they still have to be interpreted, and that means processing them through our social and political concerns. Scientists alone cannot dictate what we should do.

One might also note that many warmers are guilty of what has become known as 'environmental determinism': assuming that we can relatively unproblematically link climate change and sociopolitical change – drought leads to famine leads to breakdown, that sort of equation. Climate historians generally avoid this line of argument nowadays, and as history can attest, we can be a very adaptable species under duress – even if we fail on occasion and breakdown sometimes does occur. The interaction of humanity and the environment is a complex process that cannot be reduced to a simple cause-and-effect pattern; as we have seen, it is hard enough working out the cause-and-effect patterns in the environment alone. Brian Fagan's is a well-taken point, that the most we can say is that climate change is 'a subtle catalyst, not a cause' of significant social and political change (revolutions, collapse of cultures), as it was in the Little Ice Age period when it exacerbated tensions already latent within society.[46] Jared Diamond makes a similar point, arguing that 'there are always other contributing factors' than 'environmental damage' to societal collapse.[47] So we have to be careful about any predictions regarding 'the end of civilization as we know it', because there are still a lot of unknowns as to how things will actually develop. But one would have to say that the catalyst is likely to become progressively less subtle the further we edge up Lynas's scale, and that we too may be forced to recognize, as did our forebears, what Fagan has described as 'the brutal ties between climatic shifts and survival' in cultures under stress.[48]

Those are some of the key arguments deployed by the warmers and coolers; next up is the globalization debate, where we shall find supporters and sceptics just as much at loggerheads.

4
The Globalization Paradigm: Defenders and Detractors

Globalization has been pushed hard by market fundamentalists, and lies at the heart of International Monetary Fund (IMF) and World Bank policy. For the last few decades it has been the socio-economic paradigm, with its belief in the unfettered movement of capital, commitment to market forces dictating all national currency value, and deep antagonism towards the public sector, which it seeks to dismantle as much as possible: the less government, the better, as far as the fundamentalists are concerned. Anything that hinders the operation of the market is taken to be a social evil, and nations are effectively bullied into reconstructing their economies on the globalization model once they become the recipients of World Bank or IMF aid. Much of globalization's intellectual content is derived from the work of the American economist Milton Friedman, which will be explored later in this chapter. The impact of the latter's ideas in areas like Latin America (which became a test-bed for Friedmanite economic doctrines from the 1970s to the 1990s) has been deeply damaging socially, and the example of Argentina, which I will turn to later in the chapter as well, shows just how damaging.

Globalization is not without its critics. It has been notably attacked by Joseph Stiglitz (as in his books *Globalization and its Discontents* and *Making Globalization Work*), although ultimately he is in favour of the principle of globalization, if not the current mode of practice. More negative attacks have come from Naomi Klein, who in her recent book *The Shock Doctrine: The Rise of Disaster Capitalism*, is scathing about the globalization ethic (as she had been in her earlier book, *No Logo*, too), and in particular the ideas of Milton Friedman and the Chicago School of Economics. Stiglitz and Klein offer an impressive critique of globalization as a system, although their ultimate objectives are very different, with Klein's sympathies essentially inclining towards the anti-globalization camp, whereas Stiglitz feels globalization is necessary to the world's future well-being.

Although globalization is the current global economic paradigm, it is not always applied as rigidly by the corporate world as its leading theorists insist it should be. There can be a certain amount of licence in its application to ensure that the system works in particular countries if the local political situation demands it. China is a big enough power to be able to negotiate some concessions in this respect. The Western corporate world is so keen to have a foothold in China, arguably the most rapidly expanding market in the world today, that it will agree to some modifications to its business practices in order to accommodate Chinese sensibilities. Rupert Murdoch is a firm believer in the free market, and in the UK his papers and television channels can be very critical indeed of the government and politicians in general. Yet even he made sure that the Star cable television channel he bought in Hong Kong was careful not to broadcast any news critical of the Chinese Communist Party, after the latter put pressure on him. But there remains a bottom line: concessions such as these will only be made if they are necessary to protect profit margins. The general rule is still 'one size fits all,' and without power on the Chinese scale this is how business will proceed. I will be returning to the subject of China and globalization later in the chapter.

It would seem that cultural difference will be respected, therefore, only in so far as it serves a corporate interest. Of late, however, there has been a move amongst some major corporations to embrace what has been called corporate social responsibility, whereby companies agree to take into account the interests of the general public, wherever they are operating, as well as those of shareholders. This would seem to be an entirely commendable practice, and I will be looking into it more closely, too.

Reforming Globalization

Joseph Stiglitz's *Globalization and its Discontents* represents a stinging attack on the world economic order of the last few decades that is all the more powerful coming as it does from a one-time senior official of the World Bank (Chief Economist and Senior Vice-President), an institution that the author feels has been taken over by ideologues for whom market fundamentalism has the force of holy writ. Stiglitz runs through a list of failures that can be laid at the World Bank's door, with economic collapses in South America, the ex-Soviet empire, and south-east Asia constituting the hard evidence of the wrongheadedness of the institution's policy (as well as that of its close collaborator, the IMF). Having been set up originally to provide aid for countries recovering from the effects of the Second World War (its first name was the International Bank for Reconstruction and Development), the World Bank has subsequently been transformed into a means of ideological control on behalf of market fundamentalism. In consequence, its cures all too often have proved to be more damaging to national economies than the problems – generally either rampant inflation rates, or a collapse of the local currency on the world money markets – they were supposed to be resolving.

Stiglitz records with dismay the IMF's inability to acknowledge the failure of its policies, and dogmatic insistence that market fundamentalism was the only method that could be applied to an ailing economy. The globalization model was held to be sacrosanct, and local considerations were rarely, if ever, taken into account in its implementation. The consequences of following the IMF's dictates were often horrendous to the client countries, with sharp rises in unemployment as well as food prices pushing many of the population into poverty (just to add to the misery experienced, a recent study also indicates a distinct worsening of healthcare standards in IMF aid recipients in the wake of public sector cuts[1]). Argentina's currency and banking system collapsed, wiping out the savings of many of the country's citizens and even forcing them back to barter as the basis of economic transactions for a period. In prosperous south-east Asia various national currencies went into freefall, creating havoc on the world markets as well as in the lives of the local inhabitants. Again, unemployment and rising prices followed in their wake. The exception to the rule proved to be countries in the region where the government intervened in the market and took steps to protect its currency – Malaysia and China, neither of whom suffered to the extent of their neighbours.

Stiglitz could hardly be further from the IMF/World Bank 'tough medicine' line, putting forward the following prescription for economic health: 'Maintain the economy at as close to full employment as possible. Attaining that objective in turn, entails an expansionary (or at least not contractionary) monetary and fiscal policy, the exact mix of which would depend on the country in question.'[2] This shows a sensitivity to local conditions sadly missing in the IMF/World Bank approach, which did not materially change, even after the traumatic south-east Asia crisis. As Stiglitz emphasizes, with 'slight variants' only, the same strategy was pursued elsewhere, with the same kinds of result.[3] Discontent certainly abounded.

In *Making Globalization Work* Stiglitz is, if anything, even more critical of World Bank and IMF policies, which he sees as a major barrier to achieving anything resembling global justice when it comes to trade. American imperialism, too, comes in for harsh criticism, being pictured as unacceptably self-interested and mean-spirited, for all its grandiose claims to be setting an example to the rest of the world as to how to operate a truly democratic society for the economic benefit of all its citizens. As Will Hutton has observed, even America's many supporters would have to agree that its 'Enlightenment inheritance of free speech, free association, rule of law, free thought and pluralist checks and balances . . . today . . . look[s] somewhat tattered and even fly-blown'.[4] *Globalization and its Discontents* was designed to make us aware of why globalization was not working as far as the Third World was concerned; *Making Globalization Work* goes beyond that to outline a series of policies that

Stiglitz believes would render globalization a fairer system overall, particularly with regard to the poorer nations of the Third World who are at the moment bearing the brunt of globalization's failings. Stiglitz calls for 'a new *global social contract* between developed and developed countries', with a 'fairer trade regime'.[5] Eventually, the goal is to make the developing world economies prosperous enough to 'provide a robust market for the goods and services of the advanced industrial countries'.[6] Globalization, as Stiglitz concedes, has to date seemed 'like a pact with the devil' for most of the developing world, but he firmly believes it does not have to be that way, that the system can be reformed.[7]

Condemning Globalization

Naomi Klein has little good to say about globalization, and has conducted a spirited attack on it over a series of books, criticizing the effect it has had on the vulnerable Third World. Condemnation, rather than suggestions for reform, is more the order of the day here, Klein complaining of a situation 'where some multinationals, far from leveling the global playing field with jobs and technology for all, are in the process of mining the planet's poorest back country for unimaginable profits'.[8] *No Logo* goes after the multinationals with a vengeance, cataloguing the social ills that have followed in the wake of the systematic outsourcing of production of manufactured goods from the West to the developing world (one of globalization's most distinctive features). Klein can find little, if any, benefit that results from this process for the countries to which the multinationals have relocated production, with work practices being tolerated there that had long since been outlawed in the West. Health and safety considerations, for example, were often all but non-existent, and union activity not just frowned upon but on many occasions openly suppressed by means of armed force – often with governmental collusion. While wages stayed at barely subsistence level, the multinationals saw their profits multiply spectacularly, having cut down their production costs very significantly when compared to their Western equivalents. Some of the world's best-known brands, such as Nike and Levi-Strauss, were culpable. It is 'the international rule of the brands' that Klein is particularly incensed by, and she recommends the building up of 'a resistance – both high-tech and grassroots, both focused and fragmented – that is as global, and as capable of coordinated action, as the multinational corporations it seeks to subvert'.[9]

The Shock Doctrine is a devastating critique of the market fundamentalist cause, particularly as promoted by Milton Friedman and the Chicago School of Economics, for whom the market has an almost mystical significance that transcends politics as practised at either the national or international level. In this scheme, human beings would seem to exist to service the market rather than the other way around. Klein wades into Friedman and his acolytes, seeing them as forming what is in effect a conspiracy to force their economic theories on to the world community regardless of social or political consequences. The Friedmanites are berated for treating individual countries, such as in Latin America, as little more than 'a laissez-faire laboratory' in which they could test their theories.[10] The Friedman school, on the other hand, perceived themselves to be on a mission to correct humanity's economic mistakes (protectionism, government meddling in the market), the argument being that in the long term that was the only way to guarantee lasting economic success – conveniently forgetting Maynard Keynes's tart observation about such viewpoints that in the long term we are all dead.

The result of the spread of the market fundamentalist ethos has been a sustained campaign by the corporate sector for the privatization of as much as possible of the state's activities. If disaster strikes, as in the form of wars or natural calamities such as tsunamis, that provides a basis for economic 'shock therapy' of the type that the Chicago School favoured, and it is the corporate sector that ultimately benefits most. This is 'disaster capitalism', and for Klein it is all but a conspiracy to weaken government around the globe so that the free market can operate untrammelled.

Powerful though the argument can be, Klein's leaning towards conspiracy theory is not necessarily helpful to the anti-globalization cause. There are few shades of grey in her presentation; it is very much a case of 'them' against 'us'. She can make it sound as if everything was fine in countries like Argentina until the World Bank and IMF stepped in, and does not always acknowledge the scale of the economic problems that such countries were facing before international intervention (requested by the countries in question, we always have to remember). The status quo was patently not an option in such cases, not with the rampant inflation being experienced in Argentina anyway (running into thousands of percent annually at the worst points); somebody had to do something to amend this state of affairs. That does not, however, excuse the policies pursued by the World Bank and IMF, which generally had the effect of making a bad situation far worse. Argentina has struggled ever since to work itself clear of the socio-economic mess it ended up in, and most south-east Asian economies have had to wait several years to see their economies recover to their pre-collapse condition.

On a more optimistic note, Klein claims that the reaction to globalization by the anti-globalization movement, so called, can actually pave the way for a more meaningful form of global identity:

> At gatherings like the World Social Forum in Porto Alegre, at 'counter-summits' during World Bank meetings and on communication networks like www.tao.ca and www.indymedia.org, globalization is not restricted to a narrow series of trade and tourism transactions. It is, instead, an intricate process of thousands of people tying their destinies together simply by sharing ideas and telling stories about how abstract economic theories affect their daily lives. This movement doesn't have leaders in the traditional sense – just people determined to learn, and to pass it on.[11]

There might even be the beginnings of a 'Campaign for Real Globalization' here that would fit in with the ideas of radical democracy and cosmopolitanism that I will be exploring in more detail in Part IV.

Globalization as Robbery

Zygmunt Bauman is no more convinced of the case for market fundamentalism-led globalization than Klein is, and offers a withering assessment of its operations in his short but pithy study *Globalization*. Socially, Bauman can find almost nothing of value in the globalization process, which for this theorist is all the worse now that there is no longer any socialist bloc to offer an alternative socio-political system to the West. In this 'world without an alternative', as he dubbed it in his earlier book *Intimations of Postmodernity*, the capitalist system can proceed unchecked, and that can only be bad news for the developing world, which now has no other political force to appeal to for help as it could when the Soviet bloc existed.[12] The power unmistakably resides with the West and its market fundamentalist ethos in this new world order, where, as Bauman trenchantly puts it, '[r]obbing whole nations of their resources is called "promotion of free trade".'[13] Bauman even strikes a similar note to the apocalyptic school in global warming, claiming that, as regards the spread of globalization as a system, '*no one seems now to be in control*.'[14] From such a perspective it is difficult to see how to resist, almost as if a tipping point has been passed in terms of economic life from which there is no apparent means of recovery. Globalization has come to seem our fate; we are stuck with it, and simply have to accept the changes in behaviour and expectations that this brings about in our lives.

Bauman emphasizes the human cost of globalization, insisting that it leads to insecurity, disempowerment, and a loss of identity amongst the working population. This is seen to striking effect in the demand of so many companies for 'flexible labour':

> The pressure today is to dismantle the habits of permanent, round-the-clock, steady and regular work; what else may the slogan of 'flexible labour' mean? . . . Labour can conceivably become truly 'flexible' only if present and prospective employees lose their trained habits of day-in-day-out work, daily shifts, a permanent workplace and steady workmates' company; only if they do not become habituated to any job, and most certainly only if they abstain from (or are prevented from) developing vocational attitudes to any job currently performed and give up the morbid inclination to fantasize about job-ownership rights and responsibilities.[15]

The impact of such a policy is to render the individual all the more vulnerable, and this will be especially so in countries in the developing world, where wages are traditionally far lower and workers' rights largely notional. Flexibility can have its virtues, particularly in the professions where for many it has meant a liberation from deadening routine; but such virtues are not to be found in the lower reaches of the socio-economic system where the margins of existence are very fine. It is sad to see that the old system of life-long wage labour in boring jobs, where deadening routine is the norm, can come to seem so attractive a prospect now.

Globalization: The Human Dimension

Concerns about globalization and its effects on individuals and their local habitat also feature in Rachel Louise Snyder's *Fugitive Denim* and Fred Pearce's *Confessions of an Eco-Sinner*. Snyder looks at the globalization of the clothing industry, seeking out the stories behind those involved in the production of items such as jeans. She makes the valid point that this is an industry which does not tend to prick the public conscience in the West to quite the degree that food imports do:

> There is also lots of chatter these days about the environmental consequences of our 'food miles,' how far our fresh fruits and vegetables must travel to get to us, but 'clothes miles' are almost surely higher, given that fabric has a much longer shelf life than, say, grapes.[16]

The miles concerned become all the more complicated when you start to trace where each component of a pair of jeans comes from, as Snyder proceeds to do. The label itself is not really much help in this regard, as she points out: '"Made in Peru" might have cotton from Texas, weaving from North Carolina, cutting and sewing from Lima, washing and finishing from Mexico City, and distribution from Los Angeles.'[17] It is an example which can be replicated almost endlessly throughout the industry (and many others), with the world being treated as one large workshop: a 'borderless' world whose complex pattern of division of labour can only leave the poor consumer bewildered as to who or what she is supporting in her purchases.

Fred Pearce's concern is with his own role as a consumer within the globalization process. In an attempt to confront the guilt he feels as a serial 'eco-sinner' he travels around the world to trace where the products he regularly uses – food, clothes, or mobile phones, for example – come from, and what effect their production is having on the local way of life: the 'personal footprints' that he, like all of us, is leaving.[18] It is a depressing tale in many ways, with the underside of globalization becoming only too evident when the lives of the workers in developing countries are subjected to close examination. Wages are almost uniformly scandalously low, working and living conditions generally appalling by Western standards, child labour common (although consistently officially denied), and corruption rife through many of the industries that Pearce researches. The Bangladesh tiger prawn industry, the source of the majority of the prawn curries sold in British 'Indian' restaurants, comes off particularly badly, with Pearce being moved to cut the dish from his diet.

All of us in the West are complicit in the system and we really should be far more conscience-stricken about our collective eco-sins than we appear to be. Yet Pearce also finds some reasons for optimism, concluding that '[w]hatever the downsides of globalization, one of the upsides is that it connects us with more people from more places,' and that the more we know of others the more we can campaign to make their situation in the world trading scheme better.[19] Pearce is heartened to report that it is by no means all doom and gloom on the globalization front.

Globalization: The Answer to Poverty?

For thinkers like Martin Wolf, however, globalization is to be considered the route out of poverty, and he is one of the most passionate advocates of its virtues, which he feels are not being properly appreciated:

> The pity is not that there has been too much globalization, but that there is too little. Too many people are effectively outside the world market, largely because the jurisdictions in which they live fail to offer them and

outsiders the conditions in which productive engagement in the world economy is possible.[20]

It will only be by spreading the globalization doctrine even further that we can raise living standards for everyone, and significantly ameliorate the condition of the world's poor. In Wolf's view this means that the rest of the world should be imitating the economies of the USA and the EU, with their long-established commitment to free market principles, plus their democratic grounding. A 'liberal global economy' based on those principles is, he feels, our best defence against political totalitarianism.[21]

It is precisely the world's poorest countries who stand to benefit the most from globalization, Wolf argues, because developed economies such as America and the EU could survive reasonably well on their own if they had to, whereas in the world's smaller and less developed states there is not enough of a home market to build up and maintain a successful economy. Those who oppose such views, 'antiglobalization.com' as Wolf dubs them, are dismissed as 'utopians' living in a dream world.[22] Antiglobalization.com predictably enough includes such figures as Stiglitz and Klein, and Wolf inveighs against the efforts of that lobby in general and the 'mountainous literature of protest designed to make the reader's flesh creep' that collectively they have produced.[23] Wolf, on the other hand, is pro-globalization personified. The free market is not to be construed, as so many of its critics have it, as 'a jungle, but among the most sophisticated products of civilization', and he takes it as his brief to defend it on those uncompromising grounds.[24] There are no grey areas in this debate for this commentator; globalization leads to closer integration of the world's countries, peoples, and economies, and that is a good thing, end of the debate.

Wolf's is an intemperate argument, but so was Bauman's, and passions clearly run high on this topic. One side can see only opportunities and benefits from the globalization process, the other only exploitation and cynicism. From one perspective it is our escape route from poverty; from another, a one-way road into that condition. Wolf cannot accept that it is in anyone's interest to curtail globalization, and he has a contemptuous attitude towards those who stand in the way. It is better to be inside the globalization tent than outside, and in a phrase which can only raise the blood pressure level of most sceptics, he asserts that, '[t]he problem of the poorest is not that they are exploited, but that they are almost entirely unexploited: they live outside the world economy.'[25] Inside that economy, there is at least the chance of redressing inequality, outside it, almost none at all, and any inequality that lingers within the globalization system is a mark of political failure that cannot be laid at the door of the market. Unlike Klein and Stiglitz, Wolf thinks the power of the multinationals and the IMF is overrated, and that to give in to their critics is to raise the spectre of a return to national protectionism which would be to no one's economic or political benefit.

Economic growth clearly receives the sign of approval here, and although Wolf shows himself aware that it has the potential to create environmental problems, he thinks these are exaggerated and trusts to the good sense of the market to keep them to a minimum. The argument is that it is to a country's benefit to have a clean environment, as this is more attractive to the multinationals when they are selecting sites for production. Yet even if a country chose to host highly polluting industries, that decision could be defended, as it would keep those industries away from countries which were more environmentally sensitive. One country gets increased economic activity, the others get the chance to import the products while maintaining a clean environment at home; everyone seems to gain. The flaw in the argument is that the pollution does not just stay in the host country, it is emitted into the atmosphere, adding to its carbon dioxide levels and affecting the whole planet, environmentally sensitive and insensitive countries alike. No one can opt out of climate change; there is a globalization factor there too. Whether the market has as much good sense and as little of the jungle mentality in its make-up as Wolf seems to imply, is another open question. Klein would certainly want to press him on the assumption of a high degree of environmental consciousness among the multinationals, for example, not to mention limited power when it comes to dealing with developing nations – she seems to find the exact opposite.

The Friedman Doctrine

Next to John Maynard Keynes, Milton Friedman has been the most influential economist of modern times. Through the combined efforts of his colleagues and disciples in the Chicago School, Friedman's ideas have infiltrated all parts of the globe, and the liberalizing trend in market economics over the last few decades of the twentieth century – and on into our own – is heavily indebted to him. Friedman was a champion of liberalization and a critic of state intervention in the running of any national economy: one of the architects of what has come to be known as market fundamentalism. Friedman's dislike of the public sector took on the force of gospel for many political leaders in the later twentieth century, underpinning the policies of, for example, President Ronald Reagan and Prime Minister Margaret Thatcher in the USA and UK respectively. For leaders like these, privatization became a point of principle, which they pursued with an almost missionary zeal, pressing it on other world leaders wherever the opportunity occurred. The effect of such policies in the UK was a massive rise in unemployment, and the country experienced a considerable increase in social tension during Thatcher's period in office. It was an effect replicated in any country which was forced to adopt Friedmanite policies after applying for aid to the World Bank and the IMF, but fear of the implications of being cut off from the rest of the world economic community kept most politicians in line when it came to following the lender's prescriptions.

Friedman's magnum opus *Capitalism and Freedom* was published in 1962, and as he explained in the preface to the 1982 edition, 'its views were so far out of the mainstream that it was not reviewed by any major national publication' at the time.[26] That was to change dramatically by the 1970s, when his ideas came into vogue and were adopted by a series of national governments. Friedman's ideological creed is based on a distrust of big government and a deep belief in the virtues of the private sector. Government's major task is to ensure law and order, and, on occasion, to help initiate large-scale projects that would go beyond the means of private individuals or groups; but it should not be allowed to exceed these core functions or our freedom is put at risk:

> By relying primarily on voluntary co-operation and private enterprise, in both economic and other activities, we can insure that the private sector is a check on the powers of the governmental sector and an effective protection of freedom of speech, of religion, and of thought.[27]

While not exactly anti-government, Friedman certainly believes it is in our interests that it be kept to a minimum, insisting that '[t]he great advances of civilization . . . have never come from centralized government.'[28] At best we are to regard government as a necessary evil: a 'rule-maker and umpire', but nothing much more.[29]

Friedman puts economic freedom right at the heart of political freedom, dismissing that idea that socialism can be democratic. Real political freedom can only come through 'competitive capitalism', and this should be as unrestrained as it can be.[30] It is even contended that it is an infringement of our personal freedom for the government to force us to contribute to pension plans, so we can see just how minimal a relationship Friedman wants between government and the individual. The welfare state in general comes under attack, with Friedman arguing that a government-set minimum wage increases unemployment; that public housing creates more urban blight and broken homes; that agricultural subsidies increase the price of food. The answer is to cut back government intervention and leave it to the private and voluntary sector instead to provide these services. As for international trade, this should be released from the protectionist legislation of national states (quite a common practice in the 1960s when he was writing), including any fixing of exchange rates. A global free market would constitute 'a system of freely floating exchange rates determined in the market by private transactions without governmental intervention'.[31]

In effect, Friedman was offering us a programme for globalization, and it was those general ideas which were subsequently taken over by the IMF and the World Bank in their dealings with national economies around the globe. Many of those ideas are still informing government policy today in various countries, even if they are under a certain amount of stress owing to recent economic crises and consequent stock market volatility (a point to which I will return).

An interesting corrective to Friedman can be found in the work of Stephen A. Marglin, who in the name of community attacks the individualistic assumptions underpinning free market economics. For Marglin, that economic doctrine is destructive of some of our best impulses as human beings. Basically, what he is arguing is that, adapting Oscar Wilde, free market economists know the price of everything but the value of nothing. And by community, Marglin does not mean the state (of which he harbours some mistrust as to its authoritarian impulses), but our local communities, where traditionally there has been a strong sense of mutual obligation towards each other. The economic theories in vogue now are anathema to this kind of human relationship, preferring to see us as calculating and self-interested individuals. Hence, '[u]ndermining community is the logical and practical consequence of promoting the market system.'[32] Economics is a form of ideology for Marglin, who is deeply critical of '[t]he narrowing of the economic mind' that has occurred in recent times.[33]

The only answer is to stop thinking like an economist, which means to stop being fixated by growth and the need for individuals to accumulate more and more goods. With a nod towards the ecology lobby, Marglin encourages us to start 'living modestly' instead, rather than frittering away our natural resources as we are currently doing.[34] That would not be an end of the market as such, but it would be one way of conserving our resources for a little while longer, and according to Marglin it would improve our sense of well-being. Whether one can apply such a dictum to those living in poverty in developing nations is another matter, but at least this is one economist for whom the market is there to serve us, not the other way around.

Argentina's Economic Crises

Argentina's economic crisis in the late 1990s has been well documented. Serious inflation in the 1980s (over 3,000 per cent annually at one point) led the country to turn to the IMF for loans, with the conditions attached to these – the usual Friedmanite market fundamentalist requirements of market liberalization, a substantial reduction in the public sector, and so on – creating further problems that eventually undermined the economy. The banking system collapsed; large numbers of the population saw their savings wiped out; barter became a common method of economic transaction; unemployment soared. Further crises led to the country defaulting on its international debts in 2003, although it proceeded to recover from this situation well enough to start posting growth rates as high as 8 per cent over the next few years. Yet in 2008 it was back in difficulty again, severe enough to cause the economy minister, Martin Lousteau, to resign in frustration at the government's handling of an increasing problem with inflation. There was disagreement as to the extent of the problem, the government having adopted a new method of calculation which set it at 8.8 per cent, while critics claimed that in real terms it was about double that. Clearly, the Argentinian economy was failing to conform to the globalization model, despite years of desperately trying to do so.

The exact causes of Argentina's economic debacle are still a matter of dispute, and it has to be reiterated that the country had been mismanaged prior to applying for IMF and World Bank aid. Pegging its currency, the peso, to the dollar, as a way of bringing down inflation and stabilizing the economy, merely made the country's industries and exporters uncompetitive; when it was unpegged, the peso rapidly plummeted in value. Argentina had a long history of struggling against inflation before that, which was more to do with its internal politics than globalization as such. That means there is a danger in generalizing on this one particular example, with its own specific socio-political conditions and problems. But the fact that similar crises happened elsewhere in the aftermath of IMF/World Bank aid, with all its attached prescriptions, does suggest that the market fundamentalist, one size fits all approach is deeply flawed. What shines out in each instance is that the socio-political background was largely ignored, and that it was assumed economies could be reinvented, at one fell swoop, almost as if they were starting from ground zero: a process of 'radical erasure and creation', as

Naomi Klein has witheringly described it.[35] Economics was largely divorced from social and historical context; there was a model and it was applied without variation – when variation was what each country desperately needed. Pluralism in economic matters was not something that market fundamentalism was prepared to countenance. Globalization turned into an ideology, a grand narrative, when what was wanted was a series of narratives, each reflecting its country or region's particular history, yet still capable of interacting fruitfully with each other to mutual socio-economic advantage.

If more subtlety had been shown by the lending bodies then Argentina might not have proceeded to stagger from crisis to crisis as it has done since – and may well be in danger of continuing to do. Defenders of globalization will always claim that it is a country's fault if its prescriptions do not work, but we can see a definite pattern evolving over the last few decades in terms of the impact of coming under the IMF/World Bank wing. The need for the globalization ethos to be reformed becomes very evident, with Argentina as a frontline exhibit of how it can all go horribly wrong under the current dispensation.

Corporate Social Responsibility

The notion of corporate social responsibility (CSR) has become an issue of some note in the business community in recent years, as some of the world's larger companies have striven to improve their public image after the damage inflicted by sustained criticisms from the environmental lobby, such as Greenpeace. The implication of the fossil fuel industries in climate change is now well documented, leading them to see the virtue of developing green policies, or at least the appearance of these, in order to allay public concern and ward off any governmental intervention in their activities. Moves in this direction have to be welcomed, but critics are already claiming that in practice they amount to little more than a public relations exercise on the part of the multinationals – what has been dubbed a 'greenwash'.[36] When profits are significantly affected, for example, enthusiasm for CSR soon wanes, as the journalist Terry Macalister has reported: 'BP, meanwhile, is digging up Canadian tar sands and considering the sale of its renewable-power business. The oil group says its priority is to get its profits and share price back on track.'[37] Yet again, the interests of shareholders are taken to be dominant, and the argument that, as Macalister puts it, there should be a 'commitment to looking after "stakeholders" inside and outside a business', finds itself being cynically ignored.[38] While this is depressing news, the hope is that enough companies will see the light to continue to develop CSR further, although the problem remains that a 'business as usual' approach, particularly in the fossil fuel industry, can only serve to exacerbate global warming.

The portents, however, are not very promising. Shell, too, has signalled a shift in priorities about the renewables sector, pulling out of its involvement in the London Array wind farm, planned to be the largest offshore wind farm in the world. The oil giant complained about soaring costs for the project, despite having just posted record profits of £4 billion in the first quarter of 2008 alone. Caroline Lucas, the Green Member of the European Parliament for the south-east of England, the constituency where the farm is to be located, subsequently accused the company of 'greed' and of having 'lost its nerve and decided to shun its responsibilities in the generation of green energy'.[39] She has a point; if profits at that level are not considered enough to cushion the risk of developing renewables, then one wonders what would be. What price social conscience?

Relying on the social conscience of multinationals is no substitute for enforceable legislation it would appear, although it would help enormously if they could be encouraged to become more self-reflective about their role within society and how much good they could do if they chose. They tend to be given a rather easy ride about this by most of the general public. Discussing the notoriously poor environmental record of mining companies in Montana over the course of the twentieth century, Jared Diamond none the less is moved to remark that nothing less than toughly worded government legislation will make the mining companies act differently from how they have been doing, because 'otherwise, the companies would be operating as charities and would be violating their responsibility to their shareholders.'[40] Diamond is a sympathetic commentator, very much concerned at the damage that has been inflicted over the years on the Montana landscape he clearly loves, but it is striking how casually the assumption is made that we cannot expect business spontaneously to behave in a moral fashion – or to be capable of any charitable gesture at all. Instead, 'we the public bear the ultimate responsibility,' but that is a depressing assessment of the commercial mentality that really needs to be challenged;[41] we should not be capitulating so easily. Yet again, the shareholder explanation is trotted out, as if that absolved business from any taint of unethical conduct or from being held up as deserving of moral censure. We surely have the right to expect more than this from our fellow citizens in the business community, and we should let them know it unequivocally – while prudently also making sure that the appropriate laws are in place to ensure that they recognize our seriousness on the matter.

The Future of the Free Market

So, to return to the query we raised in Chapter 2, do we simply say that the free market is the problem, and that we should set about dismantling it as a social institution? Certainly, it cannot go on in its current form, where politicians are vying with each other to offer the biggest improvement in living standards through sustained economic growth. Neither is this purely a Western phenomenon. As John Gray has pointed out, 'China's rulers have staked everything on economic growth,' and they have taken a version of the free market model as their route to achieving this goal.[42] Since China has a population of 1.3 billion, any further significant rises in their collective living standards is highly likely to create severe pressure on the world's already dwindling energy resources. And although it is anything like evenly spread, there has been a transformation in living standards in China already over the last couple of decades, with the Communist Party's active encouragement of entrepreneurial activity opening the floodgates to rapid economic expansion, albeit under the constant monitoring of the Party bureaucracy. The profit motive is no less a factor in Chinese life nowadays than it is in the West.

Given that China is using a free market approach, if a modified one, the West will find it difficult to be overly critical of its desire to catch it up on the economic front, because that could only sound hypocritical after our development of the modern market system in the first place. The political system, however, is another matter. For commentators such as Will Hutton, this is arguably the most pressing issue that the world economic system faces at present, and he claims that, unless China moves further towards Western-style democracy and Enlightenment values, then it will not be able to maintain its current rate of growth. Were it to take Hutton's advice, then the impact this would have on global warming cannot be discounted; China's carbon emissions already exceed those of the United States (in total, if not per head of population) and to turn it into more of a Western-style society could only accelerate their growth too.

Nevertheless, Hutton is confident that it is possible to 'tilt the balance towards international collaboration' such that we can minimize the risks that China's headlong 'transition . . . to modernity' poses.[43] That does make certain assumptions about our ability to control the commercial mentality, such that it does not act in the cavalier manner that it so often espouses; no doubt the Chinese could become just as adept at playing the shareholder card as the Western multinationals have shown themselves to be. It has to be acknowledged that the Western corporate world has been notably indulgent towards China in the name of its own shareholders. Achieving commitment to CSR on both sides of this ideological divide will be no mean task, but somehow we have to start moving towards that state of affairs.

So much for the issues that are at stake. In the next section I will survey some of the major solutions that have been put forward to resolve these.

Part II

The Solutions

5
Reducing Our Carbon Footprint: Altering Lifestyles

There has been no lack of suggestions as to how we might alter our lifestyle to effect the reduction in our carbon footprint that would secure our future. Some of these are quite extreme, such as the call by radical ecologists for a massive fall in the human population globally plus a return to a simpler, agriculturally based lifestyle, as existed in pre-industrial times. Some are fairly low-profile, such as minimising our use of plastic and encouraging a far higher level of recycling in our local communities. How great an impact the latter would have on the footprint at large is debatable – which is not an argument for discontinuing the practice, as every little helps, especially if it promotes an increased awareness of what is at stake, as it seems to be doing. What would be really significant, however, is a change in our use of transport. If we could manage to cut down dramatically on car use and air and sea travel – the latter is now thought to be responsible for twice as many emissions as air travel, overturning earlier notions as to its relative cleanness – there would be a large reduction in our footprint. But that would require a principled rejection of personal car travel and overseas tourism by individuals that would be hard to engineer – never mind to monitor on a systematic basis. The arguments for and against such a rejection, where self-interest comes into open collision with the public good (differends in action yet again), will now be analysed, as will the validity of the radical ecological call for a lower global population as a means to lower overall consumption and thus the rate of carbon emissions.

Cutting back on the globalization programme is another possibility in terms of altering our lifestyle. This would mean either a drop in the volume of world trade, particularly in terms of air and sea freight, or a shift to more environmentally friendly types of transport (both airships and a return to sailing ships have been suggested in this context). The viability of such a move, which goes against the grain of our current lifestyle and the principles underpinning it, particularly the notion that the market should be left free to provide whatever the consumer happens to desire and can pay for, will be explored later.

James Lovelock, despite his apocalyptic outlook, has strongly recommended the increased use of nuclear power as an energy source, in marked contrast to the generally strident opposition to this in the Green movement. The viability of nuclear power over fossil fuel-based energy certainly needs to be reconsidered, readily available large-scale options to the latter being conspicuously thin on the ground at present, and we may well need to reassess whether we are willing to accept this as a major part of our lifestyle. Even Lovelock regards nuclear power as more of a delaying tactic to the catastrophic change in our lifestyle that will be required when the Earth's 'morbid fever' takes hold than a long-term solution. The current situation in the UK, where the government has recently announced a return to nuclear power as part of its programme for curbing carbon emissions, will form a useful little case study for examining the various arguments being aired in this debate.

Stern Words: Managing Climate Change

The *Stern Review* is a comprehensive investigation into the economics of climate change, outlining how we can set about altering our lifestyles to manage both the warming and our carbon emissions such that we can maintain something like our current way of life into the foreseeable future. Its conclusions are that we should establish an international framework for expediting the following: emissions trading, technological cooperation, action to reduce deforestation, and adaptation to the fact of climate change. None of these ideas is exactly new, but they are backed up by an impressive array of statistics as to the costs connected with either doing them or not doing them, and the project as a whole, *pace* Nigel Lawson, deserves to be taken very seriously as one of the most detailed analyses of the likely impact of global warming. For all its remarks about the need for urgency in setting up an effective international framework, the *Review* is an essentially optimistic document, arguing that it is indeed possible to move towards a lower-carbon lifestyle while still being economically successful; 'we can be "green" and grow,' it assures us.[1] In fact, if we do not become green we will cease to grow.

Stern recommends that our aim should be to achieve a stabilization of the carbon dioxide content in the atmosphere at between 450 and 550 ppm (it is currently estimated to be around 380), but that we cannot delay doing so as the process of stabilization will become progressively more difficult after that point. We have a window of opportunity we should be taking advantage of immediately. What is particularly being called for is the development of a new global consciousness on the subject of climate change, in which every country recognizes that it has a part to play and, crucially, knows that it can depend on others for help. International collective action is seen to be the key to success in improving our prospects – Ulrich Beck's 'new cosmopolitanism' and 'glocal politics' in action perhaps. Poorer countries would be financially rewarded for dropping deforestation programmes, and would also find a significant new source of income in emissions trading with the West.

The costs of achieving stabilization at the levels suggested are difficult to predict with accuracy, but Stern estimates they would probably be around 1 per cent of gross national product (GNP) in 2050 (having risen gradually up to that point). Although he concedes this is not an insignificant sum, he points out that we should remember that on current economic projections the countries in the Organization for Economic Cooperation and Development (OECD) will experience a rise in economic output of 200 per cent by then, and developing countries by as much as 400 per cent. In terms of our economic lifestyle, therefore, we are being asked to accept not so much a cut in our living standards as a slight slowdown in their rise. This seems eminently reasonable, until we reflect on how hooked we have become on rising living standards, and how unpopular politicians can find themselves when they suggest raising taxes – even for good causes. It is easier to make the argument for a change of consciousness on such matters than it is to implement the procedures it dictates as necessary; sadly, stern words are not always enough to prompt action.

The *Review* realizes that we are heading into uncharted territory when we tackle our emission levels, admitting that '[r]isk will increase along the path towards stabilization,' and that '[s]ubjective assessments have to be made where objective evidence about risks is limited, particularly those associated with more extreme climate change.'[2] But the message coming through loud and clear is that we cannot afford to do nothing at all if we want to protect our social and economic well-being. Lomborg's argument that public money would be better spent on low-profile projects is rejected on the grounds that it 'takes little account of the severe risks of very high temperature increases from climate change, which we now know are possible, or indeed likely, under business-as-usual, and which cannot be reversed if they start to appear'.[3] Here is someone who plainly does believe in fire insurance.

Saying Goodbye to Globalization

The *Stern Review* believes that the global economy can continue to expand if we take the right action, and that the private sector can play its part; but cutting back on globalization as it is practised is an option with many supporters – the anti-capitalist movement that has created havoc at several meetings of the World Trade Organization, for example. It is a seductively easy solution for critics to put forward, if an extremely difficult one to sell politically. No technological change is required, just a withdrawal from an activity, rather like quitting smoking. The problem is that most of humanity is addicted to economic progress, and can hardly conceive of a culture in which that is not the norm; modernity has dug deep within us and it will be a very hard habit indeed to break. Nevertheless, the anti-capitalist movement has generated a lot of sympathy and is likely to persist. It is also a useful corrective to some of the wilder claims made for market fundamentalism.

The radical ecology movement thinks that shrinking the population of the globe would be the surest way of resolving our climate change problems, and in an abstract sense they are right; fewer people would equal fewer carbon emissions. Globalization as we know it would most likely disappear under such a regime, and the radical ecology movement in general is not very market-oriented anyway. But in a practical sense it is hard to see how such a campaign could succeed, especially since what is often being advocated is not just a population reduction but a return to a pre-modern, agriculturally based lifestyle that hardly suits current cultural conceptions with their heavily materialist basis (we will come on to the rationale for such a policy later in the chapter). Pressure groups like EarthFirst! are in the vanguard of this movement. There is a tendency for proponents to romanticize this lifestyle as well, and although some will find their vision of a simpler, more natural existence appealing, it is unlikely to find mass acceptance. The harshness of the pre-modern lifestyle is rarely mentioned, particularly its effect on individual health.[4]

There are ways of capping the world's population, all the same, and these deserve to be explored. For one thing, the anti-contraception policies of the Catholic Church really do have to be reassessed; smaller families should become the global norm, and to demonize birth control is to prevent this happening in many of the world's most densely populated countries (in South America, Africa, and Asia, for example). Catholicism is not the only religion that frowns on birth control, and the others have to be made to realize too that it is against humankind's interests for them to continue with such a policy. Theology also will have to change in the light of global warming, and the strain this is causing on the Earth's natural resources. This point needs to be made forcefully to the world's religious leaders, because a more crowded globe is to no one's benefit; yet on current trends that is exactly what is in store for us, with better medical care improving the survival rate in large families (the norm in most of the poorer parts of the world). Estimates suggest that the world's population will increase by 2–3 billion in the next fifty years or so: a prospect which should frighten us in terms of the adverse impact it will undoubtedly have on global resources. At the very least we should be looking at ways of stabilizing global population.

Stabilizing the population was the goal set by the Zero Population Growth movement (since renamed Population Connection). This was founded in 1968 in response to the American biologist Paul Ehrlich's controversial book *The Population Bomb*, which warned, after the fashion of the nineteenth-century thinker Thomas Malthus (one of the founders of the modern apocalyptics genre), that the world's population was increasing to the point where it would soon outrun our supply of natural resources.[5] This was at a time when the world's population was only 3.5 billion, rather than the 6.7 billion it is now. In Ehrlich's view economic growth merely exacerbated the situation, making us want more and consume more. One of the movement's suggestions, as in China, is to cap families, this time at two children each. The movement has since made the connection between population and global warming, and campaigns for zero growth, or population decline, on that score.

Even China has been reconsidering its hitherto notorious 'one child per family' policy (partly on the grounds that it is extremely difficult to monitor and partly because of the abuses it has given rise to, such as female infanticide and increased abortion rates), and that could be very dangerous given its already enormous population. The policy is still officially in place, but the political will to maintain it is quite possibly weakening. Clearly, we cannot go on expanding indefinitely in population terms without exaggerating the problems we face from climate change, even if we do succeed in developing more reliable sources of green energy than at present. Population management has to become part of the response to global warming, and birth control must feature prominently in the equation. Religious ethics versus species survival is going to become an interesting debate – and it is one that cannot be delayed indefinitely.

Nature could, of course, help out in such a process of reduction, unleashing catastrophic natural disasters – volcanic or supervolcanic eruptions as cases in point – that affected climate and therefore agricultural production for the worse; or plagues of one kind or another that wiped out a significant percentage of the population (as we know happened on a frequent basis in the medieval and pre-modern world). For panicologists, AIDS is already a candidate for the latter, and in many countries of sub-Saharan Africa it is a problem of enough magnitude to inspire some doomsday predictions already. Five million or so South Africans are estimated to be HIV-positive, for example, as is a phenomenal 40 per cent of adults in neighbouring Swaziland (the highest incidence in the world). Global warming is also likely to encourage the development of many tropical diseases, although whether it turns out to be on a scale that would seriously affect the world's population levels remains to be seen. Many tropical diseases are already moving into higher latitudes, however, and the more the ecosystem is disturbed by warming, then the more vulnerable it becomes to the spread of plant diseases too, bringing agricultural problems in their wake. These are hardly factors we would want to build into a planned campaign of action to counter climate change, although no doubt some would see any increase in disease as a case of Gaia taking a just revenge on humanity for its presumption. It is to be hoped this is not the way the global population declines, but it is a possible outcome unless we change our ways.

Going Pre-Industrial

So what are the arguments that radical ecologists put forward for a return to a pre-industrial lifestyle? At base, it is a case of finite resources; we cannot go on as we are because the Earth simply cannot sustain it. Peak oil is a highly symbolic warning for us, and its ramifications are as yet not well recognized; car production continues apace, and car ownership to grow. Food production too cannot go on increasing indefinitely to cope with an ever-expanding global population - and never mind an expanding population, what with droughts, floods, and the shift from food crops into biofuels, it is going to become increasingly difficult to feed even the existing population. A host of other resources are under stress globally – water supplies being well to the fore, with many parts of the world struggling to meet demand as it is. As we saw from the International Alert report (see Chapter 1), such shortages are becoming an increasing cause of conflict, especially in the world's poorer areas, and there is every indication that the situation is likely to get progressively worse. None of those resources is likely to increase – we may well have to face the prospect of peak water soon, for instance – and that is all the more reason to take the radical ecology line seriously, even if most of us will probably want to stop short of their preferred solution of a return to the pre-industrial lifestyle.

There is also an ideological line being peddled here, which takes a very different view of humanity and its relation to nature and other species than has been the norm in modern times. Modernity encouraged us to exploit nature and the animal world for our benefit, and that has been the driving force behind the development of the free market system and the globalization ethic; but radical ecologists ask us to renounce this outlook and rein in our wants and desires for the good of the planet. They call for us to be much more modest as a species and to drop our assumption of superiority; we are only one species amongst many and have no divine right to dominate our world. Living in harmony with our environment is what they argue we should be striving to achieve instead – and if we do not manage to do this, disaster is waiting in the wings. As the deep ecology theorists Bill Devall and George Sessions put it, the goal is 'a new balance and harmony between individuals, communities and all of Nature', and this means that '[w]e must take direct action.'[6]

EarthFirst!, as their name alone suggests, take a much more aggressive line concerning the human–environment relationship, and really do want to put the clock back to an earlier era of our development. One of their key slogans is 'Back to the Pleistocene,' and one can hardly be blunter than that in political terms;[7] what could rising economies like China and India possibly make of such a call? EarthFirst! certainly do favour direct action as well, up to the level of violent response on occasion (they have been known to set explosions in tourist facilities), and nuclear power is nowhere on their agenda. If we were to be projected back to a Pleistocene lifestyle by events in the natural world (something on the unlikely, but not impossible, lines of a meteorite strike of the kind that has been theorized to have wiped out the dinosaurs), then we would have to deal with it somehow, but one cannot imagine this happening on a voluntary basis – not en masse, anyway.

Transport: To Fly or Not to Fly?

Air flights currently account for 1.6 per cent of our total carbon emissions annually, but air travel has been a rapidly growing market for some time now and looks set to continue that way. The British government has estimated that the number of airline passengers will more than double from 228 million in 2005 to 480 million in 2030. That figure relates to British passengers only; presumably we can expect comparable increases elsewhere around the world, which is a daunting prospect. In George Monbiot's rueful assessment, '[b]efore long, there will scarcely be a patch of sky without a jet in it.'[8] And if those jets continue to use fossil fuel, that will mean they will be releasing double the amount of carbon dioxide into the atmosphere that they are now, with the added problem that emissions that occur at high altitude are thought to have more effect than if released at ground level. The *Stern Review* points out that '[t]he uncertainties over the overall impact of aviation on climate change mean that there is currently no internationally recognised method of converting CO_2 emissions into the full CO_2 equivalent quantity':[9] yet another worrying unknown over which we should ponder.

Air freight is big business too, and as long as we are hooked on out-of-season fruits and baby vegetables it is likely to remain so. We can expect more jets in the sky from that quarter too, criss-crossing the globe relentlessly on the Western consumer's behalf. While airlines are exploring the possibility of using other fuels (including biofuels), and also of improving the design of aircraft such that they burn fuel more efficiently, as yet there has been little real progress made and we look to be facing yet another crisis on the emission front. Green air travel is at the moment no more than a distant hope.

Hydrogen continues to attract a lot of interest as an alternative source of fuel for both planes and cars, but the technology is still very underdeveloped. The main drawback is that hydrogen is not as efficient as standard fossil fuels and requires very large storage tanks, thus cutting passenger room considerably. There is also the problem that hydrogen-powered planes demand a different design which would require them to fly at higher altitudes than jets currently do. There, the water vapour that hydrogen fuel gives off would become a problem, as at such altitudes water vapour counts as a greenhouse gas. The Royal Commission on Environmental Pollution has estimated that such emissions would be as much as thirteen times that of fossil fuel-powered aircraft.[10] Improvements may well be forthcoming, but hydrogen clearly is not an answer as yet.

The revival of airships is periodically suggested as a means to cut down on the volume of standard air flights, and a few companies have entered the market with this in mind. The ships could be powered by a combination of helium and hydrogen, and the claim is that they would be far safer than their earlier versions, which fell into disrepute after several large-scale disasters such as the loss by fire of the Hindenburg in 1937. Travel times would be significantly slower than by plane – 43 hours to cross the Atlantic, for example – but that will be something we must learn to expect in a world trying to come to terms with emission reduction while still maintaining some semblance of a system of mass transport. Putting the positive case, George Monbiot points out that airships would be much more spacious for passengers than standard air travel, so that they could start thinking of journeys by them as 'rather like travelling by cruise ship, but at twice the speed and using a fraction of the fuel'.[11]

Airships are certainly far cleaner than conventional air travel, and even the water vapour emitted by hydrogen has little effect at their much lower flying altitudes (usually around 4,000 feet). But they seem to have had little impact on government thinking to date, and one would have to say that the journey times would be regarded as a drawback by most for long-haul travel. Quick weekend breaks to exotic locations would become a thing of the past; a weekend in New York for Christmas shopping has become quite popular with the British public in recent years, for example, but it could hardly survive a 43-hour trip each way. The alteration in our lifestyles is possible, however, but as Monbiot warns, there are resource problems with airships no less than with fossil fuel-based transport. Global supplies of helium are estimated at only around 50 years, but of course airship travel would cut that back considerably. After peak oil we could find ourselves facing peak helium. The only safe way to green the air looks to be to fly far less often.

Transport: The Future of Cars

Modern society has such a love affair with the car that it is almost impossible to envisage living without it. Put simply, cars are a symbol of economic success. It is mark of any growing national economy, therefore, that car ownership increases dramatically, and we can only observe the rapidly expanding economies of India and China with trepidation in this respect. India has just unveiled a new cheap car (the Tato Nano) designed for the mass market that is fast emerging in the country due to its economic boom, and if this proves to be as popular as is hoped by its manufacturers then we can look forward to a significant increase in carbon emissions from a growing car-owning constituency in the subcontinent. Clearly, this will not help in the effort to combat climate change, although it does raise the thorny issue again of how a developed West can argue with any credibility against development elsewhere.

There are more options with cars than with planes, however, and various other fuels are viable. Biofuel is the front-runner at present, and hybrid cars making use of this are already on the market. But there are potentially severe drawbacks with biofuel, as we shall go on to discuss in the next chapter; it may, indirectly admittedly, increase carbon emissions, and it is already beginning to have a detrimental effect on food production and costs. Hydrogen is a possibility, although, as we noted above with regard to planes, it does demand increased storage capacity; that is less of a problem with ground travel than in the air, however, where regular refuelling is not on offer. There are other modes of transport available to us as well, such as trains and buses – even bicycles. Massively increased investment in public transport could lead to less dependence on the car; but the record on this around the world at the moment is, at best, patchy. Many developing countries have no mass public transport systems at all. There really does need to be a shift away from cars to other types of transport, particularly for short-haul local journeys, but that is a lifestyle issue that most of the population has not yet addressed as seriously as they should; we remain, as Lynn Sloman has put it, 'car sick'.[12] It may well be that peak oil could push fuel prices up to such an extent that car usage declines, but that would put pressure on politicians to reduce fuel taxes so the outcome is not entirely predictable on that score either – although it will be an interesting development to watch.

Sea Change

Sea travel was the norm for long-distance travel, and trade, until the invention of the aeroplane, and it might be thought that it is due for a renaissance if air travel and air freight are becoming so problematical. Unfortunately, this is yet another case where we are going to have to countenance a change of lifestyle, because it now seems that the carbon emissions from shipping have been drastically underestimated and are at a level that should be worrying us. Recent research records that global shipping's contribution to carbon emissions is far higher than air traffic's, 4.5 per cent compared to 1.6 per cent, and the fact that these emissions have often not been included in national totals of emissions is a scandal that has only just come to light.

Greening the shipping world is another possibility, however, and one that is beginning to attract interest. Again, this is a matter of reassessing the types of fuel used and the actual construction of the boats themselves. Solar power has been suggested as a solution to the fuel problem, and hydrogen has been mooted as a possibility too. More streamlined designs for the boats would cut fuel consumption, as would travelling at slower speeds than usual (the same applies to cars). A more radical suggestion is to return to sailing ships, which do not require fossil fuels at all. Technology has developed to the extent that these can be made far more sophisticated than their predecessors, and could be built on the substantial scale that modern trading requires. Large 'kite' sails could be used to maximize the effect of the wind. Speeds would be much slower than we are used to, but we can surely encompass that in our lifestyle if the environmental benefits are clear.

'Super-size Me': The Revival of Nuclear

If we do not wish to return to the past, perhaps we can return to what used to be considered our bright and shining future – nuclear power. It is now official government policy in the UK to expand its network of nuclear power stations, after a spell in which this facility had been allowed to run down because of public fears about its safety. Periodic accidents, as well as health studies indicating a greater incidence of various forms of cancer in the areas around power stations, had led to a climate opposed to nuclear power, particularly when Britain was starting to reap the benefits of its own oil fields in the North Sea in the later twentieth century.[13] Nuclear power began to seem unnecessary, certainly unpopular (always a consideration with a government, wary of alienating the electorate), and not worth the health hazard it posed – a hazard that became all the more real after the disaster at Chernobyl, when the latent dangers registered very forcefully throughout Western Europe. Chernobyl had considerable symbolic significance, and it was taken to spell out a warning as to the unreliability of the nuclear power option. Why take that chance when oil, and increasingly natural gas, were so freely available? Other countries still went ahead with their nuclear programme (France most notably, which by 2008 was receiving 79 per cent of its electricity from nuclear power), but the overall popularity of the nuclear option clearly waned for quite a while, becoming 'dead on its feet', as one commentator has summed it up.[14]

There was also always the fear that non-Western countries who continued to develop nuclear power (North Korea, for example, and then latterly Iran) were doing so more with weapon production in mind than safeguarding their energy needs. Predictably, the West was quick to condemn such development, which merely added to the air of unease surrounding nuclear power in general. How the West will reconcile this suspicion of Iran in particular with its own renewed commitment to nuclear-derived energy remains to be seen; it can only be perceived as a case of 'do as I say, not as I do,' with all the colonialist overtones such an attitude inevitably carries. Iran has spent very heavily on developing its nuclear power capability, and has been adamant that it is for peaceful purposes only. Other such clashes of interest are only too likely to occur in the future with nations such as Pakistan, whose nuclear capability would be a real concern were it to reconsider its generally pro-Western stance (a not impossible occurrence given the powerful Islamist elements active within Pakistani politics). Some national lifestyle decisions will be very hard to alter, one might conclude.

Then there is the problem with radioactive waste, which has bedevilled all nations to have gone down the nuclear route. Radioactive waste is extremely potent, some of its isotopes having a half-life of millions or even billions of years, so we have to acknowledge that we are passing on problems to a whole series of future generations – perhaps altering their lifestyles for the worse. No really satisfactory solution to the disposal of this waste has been found to date, and although various things have been tried, their long-term effects remain unknown. This is yet another area in which we must 'expect the unexpected', as William Laurance has warned us in dealing with non-linear systems such as the environment is made up of, since, as we have found out already to our cost, 'unknown unknowns are much more likely to spring surprises when a system is stressed.'[15] It is hard to see how dumping ever-increasing quantities of radioactive waste into the environment would not qualify as stressing the system unduly.

Nevertheless, the UK, in partnership with France, has now embarked on an ambitious programme to build several new 'super-sized' nuclear power plants around England, with the government's business secretary, John Hutton, declaring himself committed to increasing the amount of electricity generated from nuclear power to 'significantly above the current level' of 19 per cent.[16] The mood is one of unabashed optimism, with the British and French governments convinced that nuclear power will, in the words of an industry consultant, provide the means to 'save the world'[17]: just the kind of thing with which politicians want to be identified. John Hutton has pressed the economic button with this issue to make it sound even more attractive to both the public and the political class, arguing that Britain can be 'the gateway to a new nuclear renaissance across Europe', and that this will lead to the creation of 100,000 new jobs.[18] This is one case at least when the establishment is quite happy to heed the advice of our most apocalyptic thinker, James Lovelock. Like it or not, and the Green movement patently does not, nuclear is set to become a major player in energy production in the UK: in many ways, the easy option for a government to choose, especially when it can claim that its resurgence will lead to an economic revival as well. The public's green sympathies rather notoriously tend to slip when it comes to the economy. Back to the future with nuclear seems to be the current thinking, and the public appears to have accepted this as the lesser of several evils and to be prepared to take its chances.

The momentum behind the revival of nuclear is building up remorselessly, and its supporters are adamant that this is the only sensible way forward at present while we do our best to develop longer-term strategies. In the assessment of the American science writer William Calvin, for example, 'that means going with what we've got, the current approved reactor designs. I'd prefer deep geothermal heat if they can ramp it up fast enough. But those are the only two routes, so far as I can see, likely to create our safety margin during the next decade'[19] (deep geothermal heat involves drilling down into hot, dry rocks, piping in water, and then forcing steam to the surface to drive turbines). Meanwhile, we will just have to hope that an effective method of disposing of the vastly increased amount of radioactive waste that will accrue from the new generation of super-sized nuclear power stations is soon found; otherwise we face the prospect of a potentially badly contaminated planet seriously injurious to human health. None of the current suggested methods of disposal, such as burying the waste deep underground or under the seabed, is guaranteed foolproof, and if leaks were to occur, they could work their way into our water supplies or the food chain with disastrous consequences. We might resolve our present energy needs only to pass on yet another intractable environmental problem to future generations.

Increased carbon emissions or nuclear contamination is not a particularly attractive choice to have to make – and it is interesting to note that the waste problem has just been passed over in the current official campaign for nuclear power. We are told the advantages of a return to nuclear, but none of the potential disadvantages (expected or not). It has to be conceded that there simply is no completely unproblematical way of continuing to use energy at our present rate. If nuclear is the lesser of several evils, then it is still an evil of some risk to us, not to mention the many generations into the future saddled with the increasing quantities of waste, and should be acknowledged as such by all the parties concerned.

Surviving Green

Nuclear power is a high-profile solution to altering our lifestyle so as to combat global warming, and a very public one too, but lower-key, more personal approaches have also been recommended. We all know about the virtues of recycling and reducing our use of plastic bags, bottles, and paper, but there is also Brian Clegg's survival manual for the climate change age to consider. Clegg sets out to offer 'clear-headed, practical guidance so that you, your family and loved ones can prepare for the end of the world as we know it'.[20] The author assumes that we are facing something like a return to the state of nature, where every family will be engaged in a desperate struggle to maximize their own resources in order to survive. This will be a world where power sources regularly fail, and humanity will divide into two groups: 'the majority of the population will take astonishingly little action to protect themselves, even in the face of the starkest possible warnings . . . Some, however, will take heed and prepare accordingly. Forewarned is forearmed.'[21] It is not a very pleasant vision of what our future might be, and one suspects that the selfish behaviour it is encouraging would lead to social conflict, as has happened in the past in war-torn nations, with citizen turning against fellow citizen when order breaks down.

For Fred Pearce this is 'the new green survivalism', and he finds it both depressing and short-sighted:[22] more of a symptom of the problem we face than any truly meaningful solution to it that should be widely adopted. I would hope that most of us would see this as a case of altering our lifestyle for the worse in response to climate change, and that we can resist the urge to withdraw into ourselves and our own personal interests in this manner. Green survivalism looks like a regressive step in terms of the carbon footprint wars, although one suspects we are due to hear more from it if the situation starts deteriorating rapidly and individuals feel thrown back on their own devices. There is always the chance that technology might save us, however, and I will turn next to some of the schemes that have been devised with that objective in mind.

6
Living With Our Carbon Footprint: The Technological Response

Altering lifestyles is a notoriously difficult task, and many scientists and politicians have put their faith instead in technological solutions to global warming. Both President George W. Bush and Prime Minister Tony Blair, for example, expressed such faith during their terms in office in an attempt to allay growing public concern about the lack of governmental action on the carbon emission front. The message was that we could rely on scientific ingenuity to arrest the slide towards tipping points – very much the modernist approach to environmental problems: seek a better technology. Thus biofuel is increasingly touted as an alternative to purely fossil-based fuels; wind and solar farms are advocated as alternative sources of energy; even nuclear power has found a new group of champions, as can be seen from the arguments of James Lovelock and the current British government. Bjorn Lomborg's proposals for containing global warming by a significant investment in more environmentally friendly technology also need to be considered in this context, to discover just what it is that the sceptics find so congenial about his analyses.

More extreme solutions such as geoengineering are also being promoted by the scientific community, on the assumption that something more drastic than incremental technological change might be required if climate change comes to be seriously out of control. Geoengineering involves some cutting-edge science, and various proposals for undertaking it will be surveyed, such as placing mirrors in orbit round the Earth to alter the amount of solar radiation we receive, or engaging in iron fertilization of the oceans to increase the growth of algae, which are acknowledged to be highly effective at absorbing carbon. It is not always the technical feasibility of such processes that are at issue, however (although these can sometimes stretch credibility more than a little), so much as the after-effects of their implementation. At that point we move into uncharted territory, potentially very dangerous uncharted territory where the amplifying effects of schemes gone wrong could be catastrophic for both the environment and humanity. Once we slide into positive feedback we can rapidly become helpless, and as long as that remains a significant threat we have a right to demand caution on the part of the relevant authorities.

The Biofuel Solution

Biofuels are being pushed quite hard by various governments around the globe. The EU, for example, is requiring all countries under its jurisdiction to ensure that 5.75 per cent of the petrol and diesel it uses is obtained from renewable sources such as biofuel by 2010, and 10 per cent by 2020 (although that policy may be reconsidered, as the EU is beginning to show some concern about the impact of biofuel production on food prices). The Renewable Transport Fuels Obligation has meant that, as of 1 April 2008, all petrol and diesel sold in the UK must include 2.5 per cent of biofuel. There have, however, been complaints from leading scientists that biofuel is the wrong road to take to carbon reduction, and that its effect may even be to increase emissions overall once factors such as deforestation are taken into account. Professor Robert Watson, the British government's chief environmental scientist, has gone so far as to call the move over to biofuels 'insane'.[1] Nevertheless, the investment in biofuel production is now so substantial that it will be very difficult to run down, never mind suspend altogether as critics such as Professor Watson are openly advocating. Not for the first time in this area, the cure is turning out to accelerate the disease.

Neither is it exactly helpful to the reduction cause that some supposedly 'green energy' companies in the UK have discovered that by importing biofuel to the USA and adding some locally produced biofuel to it there, they can then claim an American subsidy of 11p a litre and ship it back to Europe to sell as an imported product at below current rates for biofuel on this side of the Atlantic. This so-called 'splash and dash' technique has been attacked in the press, who have dubbed it a 'scam', which does not seem unreasonable; but it is at the moment legal, even though, as one reporter has observed, it certainly 'flouts the spirit of producing green fuel by transporting it needlessly across the Atlantic at a time when campaigners are voicing concern about emissions from global shipping'.[2] Once again, we find that by leaving energy to the free market we leave ourselves open to sharp practices which have profit only in mind, and to cavalier traders for whom searching for loopholes in legislation is a way of life. As a *Guardian* editorial remarked, what the market demonstrates at such points is a 'breathtaking cynicism', but until the current ethos of the lightest possible regulation on business is reassessed, that is what we shall have to cope with – and to our collective detriment.[3]

The Mirror Solution

So, what would happen if we placed mirrors in orbit with the objective of lessening the solar radiation count? The theory is that each mirror would reflect the radiation back into space rather than it reaching Earth and being radiated back into the atmosphere, to be trapped by the greenhouse gases, and that if this could be done on a sufficient scale it would have a significant effect on the global climate. In one scheme the mirrors would be very large indeed, as much as 1,000 kilometres in diameter, and located far out in space at the Lagrange point between the Earth and the sun (where the forces of the two bodies cancel each other out, enabling an object such as a mirror to remain in orbit there). Another scheme devised by the astronomer Roger Angel calls instead for 16 trillion small discs, each about 2 feet wide and weighing only a gram, to be placed at the same point; although it would have to be said that the logistics to this are somewhat mind-bending, requiring twenty launches at a time every 5 minutes for 10 years.[4] Then there is the cost: $5 trillion. A similar effect might be achieved by releasing billions of reflective balloons into the atmosphere. Climate modelling suggests that as little as 1.8 per cent reduction in solar radiation would offset a rise in carbon dioxide levels up to 550 ppm, thus preventing further temperature rises for the time being.

The American National Academy of Sciences has reckoned that if we had 55,000 mirrors 100 square kilometres in size orbiting the Earth, enough sunlight could be reflected back into space to reduce our carbon dioxide levels by about a half of present levels. That would be a huge undertaking, at enormous cost, but it sounds technologically plausible. One journalist has described such schemes as our 'plan B' if all else fails, therefore worth scientists' continued thought experiments, wildly eccentric though they may seem much of the time.[5]

Some scientists have been inspired to take the mirror notion even further, suggesting that we put them on the moon. Initially, the idea was to help us announce our presence to alien life forms by increasing the light sent into space from the Earth-moon system, flashing it in regular patterns to indicate its deliberately engineered source. To achieve a 20% increase in that light emission the moon would need to have half of its surface covered by mirrors. If those mirrors had photovoltaic cells fitted to their undersides they could also manufacture electricity, which could then be beamed down to Earth by means of microwaves. One of the proponents of the idea, Shawn Domagal-Goldman, has proudly claimed that as well as making us more visible to aliens, the scheme therefore 'could help solve the climate crisis, too'.[6] There is no denying the amount of ingenuity that is going into providing solutions for climate change, although in this case it would have to be said that the practicality of the projected scheme is another matter.

A similar idea to the mirrors was the 'sulphur sunshade' mooted by the physicist Edward Teller in 1998, whereby sulphate particles would be pumped into the atmosphere such that they created the effect of a sunshield, thus keeping out enough solar radiation to cool the planet down. The notion came from the observation that the sulphur released into the atmosphere by volcanic eruptions reduced global temperatures – often for quite a long period afterwards ('the year without a summer'). Such an effect could be replicated by having naval guns fire sulphur pellets into the atmosphere; on a large enough scale, according to proponents, this ought to be enough to counter the warming trend.

The Albedo Enhancement Solution

Nevertheless, even scientists in favour of geoengineering can have very strong reservations about its ethical aspects: 'I don't see how you decide on the basis of all humanity how to change the planet. But I think it's irresponsible, in a way, not to study it,' being a not untypical response from a leading climate researcher.[7] Some of their peers who are even more sceptical of the enterprise in general have therefore suggested that we opt for simpler measures, such as enhancing the Earth's albedo – that is, its reflectivity – on its surface. Alvia Gaskill, for example, has put forward an ambitious programme entitled the 'Global Albedo Enhancement Project' (GAEP) that would lead to more solar radiation being reflected back into the atmosphere, thus cooling the surface of the planet to a noticeable degree. It is well known that the polar areas reflect more than dense forests or deserts, for example; hence the concern being expressed about the progressive shrinking of the former (the principle is the same as wearing white rather than black clothing in the summer).

Albedo enhancement has been described by Gaskill as a 'delaying tactic' to slow temperature rises over the next few decades, while in the interim, it is hoped, more effective technologies for dealing with climate change can be developed.[8] It involves some very simple measures, such as the whitening of pavements and roads in urban areas, which are known to absorb more heat than the countryside (the urban heat island effect, as this has been called). Gaskill claims that if 2,500 square kilometres of the pavements and roofs in Los Angeles – constituting 15 per cent of the city's surface area – had their albedo raised by 7.5 per cent in this manner, and an extra 10 million trees were planted to reinforce this scheme, then the city's temperature could be reduced by as much as 5°F overall. The benefits of such a project would be substantial: 'This would greatly reduce the demand for electricity to power air conditioners as well as reduce the formation of ground level ozone that increases with increasing temperature.'[9]

While whitening large areas of cities like Los Angeles may sound quirky, it is not altogether unimaginable. After all, we are used to the sight of 'white' villages in the Mediterranean area, where the same effect is being sought by the inhabitants; we even find these picturesque and aesthetically pleasing when we visit places like the Greek islands. But Gaskill is thinking on an even bigger scale than this: to covering over entire desert areas, such as the Sahara, the Arabian, and the Gobi, with white plastic sheeting for a period of up to 60 years. If modelling and field testing yield promising results then she recommends we start such a programme in 2010, and continue it until 2070 or so, proceeding at the rate of 67,000 square miles of coverage annually. The technology is straightforward, and the cost relatively low compared to the more dramatic mirror projects (the first year of coverage being estimated at around $14 billion).

Gaskill admits that an enhancement programme of the kind she is proposing would not be as efficient as preventing the solar radiation from entering the atmosphere in the first place (as in the mirror method), since only half of that radiation then proceeds to reach the surface of the Earth; but she is confident that savings can be made from the still substantial total. Enhancement is more practical than any of the geoengineering proposals we have just considered – certainly more so than the prospect of delivering 16 trillion discs to the Lagrange point and keeping them aligned correctly there.[10] Yet, even if it is somewhat on the eccentric side, it does lack the sheer daring and headline-grabbing potential of those other projects. Just as apocalyptics fascinate us, so do grandiose schemes about engineering the operation of our galaxy to suit our own needs. Geoengineering into deep space has the effect of turning climate change into an exciting adventure – although whether that is a good thing is debatable, as it tends to direct our attention away from the more practical business of altering the lifestyle that landed us with the problem in the first place.

The Algae Solution

Increasing the growth of algae in the ocean – the ideal site is thought to be the lower reaches of the southern ocean going down to Antarctica – is another process that is well within current technological capabilities and easy enough in principle to implement. The science behind it also seems very straightforward and much easier to monitor. Algae is well known for its capacity to absorb carbon dioxide, and iron fertilization is the obvious way to increase its growth. When the algae die they sink to the ocean bed, taking the carbon they have absorbed with them and thus removing a substantial amount from circulation. The iron could be dumped into the ocean directly from large ships such as supertankers, helping to stimulate the growth of the algae. The cost of introducing enough iron to make a noticeable difference in our greenhouse gas production has been estimated at $100 billion, which makes it one of the cheaper methods on offer.

A more ingenious method of encouraging algae to grow has been proposed by James Lovelock and Chris Rapley (the director of the Science Museum in London). This is to pump up large volumes of cold water from the seabed, in pipes 100–200 metres long and 10 metres in diameter, to encourage surface algae to grow. By the standards of geoengineering in general such a project would involve a modest amount of material, although the authors are careful to warn that 'the impact on ocean acidification will need to be taken into account.'[11]

Yet another role for algae has been proposed: to absorb carbon dioxide from power stations so that it is much cleaner when it is emitted (estimates suggest by 40 per cent or so). In this scheme tubes of algae are attached to the smoke stacks and act as a sponge, soaking up the exhaust gases from the plant. Several power companies in America are already trialling the method.

Separating the Wood from the Trees: The Biomass Solution

Solutions so often seem to have adverse side effects that one can begin to despair, but scientific ingenuity remains undaunted, as we can see in the proposal from the American atmospheric scientist Nin Zeng to bury trees underground as a means of preventing an important source of carbon from being released into the atmosphere.[12] As the science writer Richard Lovett has remarked, Zeng had noted that landfill sites in America 'were acting like carbon sinks', and this gave him the idea of burying trees such that they could not release carbon as they aged (decaying trees give off more carbon than they absorb).[13] Zeng argued that if half of the wood that grew in the world each year was buried, an activity that would involve a global workforce of around a million, then we could effectively offset our current annual fossil fuel emissions. It would be an arduous task; forests would have to be thinned on a regular basis to make space for the burial and the trees would have to be buried quite deeply – between 5 and 20 metres – to prevent the carbon escaping after being broken down by natural processes. An alternative would be to store the dead wood in overground shelters, but that seems a far less feasible, and certainly much more obtrusive solution. Burial offers the prospect of a long-term 'lockdown' of the carbon – between 100 and 1,000 years Zeng has estimated. In similar vein, other researchers have suggested that the restoration of natural carbon sinks, such as peat bogs and marshes, should be made a priority.

If the practicality of Zeng's proposal is questionable (although not impossible if the political will were there to initiate such a programme), then the claims made for its efficacy are even more so. The process would only work if the conditions were such that the wood did not start to rot – once it did, carbon would find its way back out into the atmosphere again. The type of soil the wood is buried in would also be crucial; soil with a high population of termites would not work, for example, because the termites eventually would eat their way through the wood and then release the carbon through their bodies. Other types of soil might give off methane, which as we have seen, is one of the most potent of greenhouse gases, and one that we should be striving to minimize wherever possible. Then there is the problem, as with nuclear waste, that we are solving a problem now at the expense of generations to come – in this case, potentially in as little as a century or so, when Zeng's buried wood might indeed be beginning to rot (if not devoured by termites before then). There is no denying that it is an intriguing idea, non-linear thinking by any standards, and that it may well come to be refined to the point where it seems worth a try. At the very least it is technically straightforward; just dig a lot of holes in the ground, tip the trees in, and then cover them up again. But yet again we find ourselves running into what we might start to call the law of unintended consequences: that there will always be such, no matter what the scheme being floated, and that they will probably be to your disadvantage – sooner or later.

Nevertheless, there is a growing campaign to encourage the lockdown of carbon in biomass. The American soil geochemist Jim Amonette has even suggested that landfills could become a method of obtaining carbon credits. While admitting that the biomass solution in general leaves many unresolved issues (as also with burning it, about which see below), Amonette feels it is still well worth developing, on the grounds that 'it's better than the disaster of waiting 40 or 50 years for the perfect solution to be found.'[14] He has a point, and the note of urgency is justified; but it does depend how realistic the trade-off is going to be with the projected unintended consequences. We are always in uncharted territory when faced with such calculations. And do we really want to see landfills proliferating across the countryside, with logging companies and their ilk seeking to absolve themselves of their environmentally unsound practices elsewhere by building up credits there? The carbon credit strategy, as we have seen, is rarely that straightforward.

Renewable Power Sources

Solar power, wind power, wave power: these are three key sources of renewable power which have been earmarked for development. The planet is certainly not short of any of them, although they require a fair amount of work to render them commercially viable on the scale that is needed. Most countries in the West have invested in at least some of these, and in some cases, as in Germany, they have become a very significant element in the government's energy planning. Claims have even been made that a combination of solar, wind, biomass, and geothermal-derived power could provide 90 per cent of the USA's energy needs by 2100, including all of the country's electricity, although this would involve a huge investment from the federal government – more than $400 billion in the first instance just to get the solar power scheme, which would be the key component, under way.[15] One of the justifications put forward for such substantial investment is that the development of renewables on a major scale would reduce America's imports of oil, thus making significant savings that would offset the admittedly high cost. In Germany that message appears to have got through to the government already.

Solar radiation is the largest source of renewable energy that we have, and it is the subject of intense technological development at present, as in the area of the production of photovoltaic cells. The cheaper these can become, then the more viable that solar energy will be as a mass source of power, and costs are beginning to drop markedly in the fight to corner the potentially huge market that could exist. Sunlight is not entirely predictable, and it is spread unevenly across the globe, but in theory it could provide the greater part of our energy needs if we can just get the technology right. That is exactly the kind of challenge the scientific and technological communities like, and they are responding to it with enthusiasm.

The technology to harness solar power is well understood, and it is already in operation in many countries. In the American southwest it is attracting considerable commercial investment, and various methods are being used: parabolic trough technologies, central receiver systems, dish systems, and concentrating photovoltaic systems (CPV), for example. In the first two cases, the sun is tracked with mirrors, which heat up a fluid to produce steam to drive a turbine. Dish systems involve concentrating sunlight on to a parabolic dish which heats a receiver that then powers a thermal engine. CPV uses mirrors that track the sun to focus the light on to solar cells. Parabolic trough technology is currently the most efficient method, but CPV has a lot of proponents, and can also be installed on rooftops, thus allowing homeowners to generate a substantial amount of their own electricity. This would lessen demand from the plants.

Solar power works best in climates like the American south-west, but even in less sunny areas it can still be successful. Germany has devoted considerable resource to developing this area, and has turned into a world leader in this respect, producing around a half of the world's entire total of solar-generated electricity in 2006. The country is also a leading producer of solar panels and photovoltaic cells, which it sells on the world market. The plan is to provide 3 per cent of the nation's energy needs through solar power by 2012, with home production (rooftop CPV) also being actively encouraged by the government. Homeowners can even sell any surplus electricity they produce to the national grid, and are paid at premium rates when they do. All this in a country which receives only about half the sunshine of its Southern European neighbours, never mind the baking American south-west. Overall, Germany's intention is to achieve a quarter of its energy supply from renewables by 2020.

Since one of the effects of rising oil prices and the prospect of peak oil has been to revive interest in coal production in some countries (Japan, for example[16]), it sounds worth persevering with solar power. Coal is certainly not carbon-neutral, and it would be a retrograde step to go back to it on any large scale – especially now that we know the role it plays in global dimming. Environmentalists will continue to press the case for greater development of renewables, but coal is a more reliable source of power overall compared to most renewables (certainly to wind and wave), and the temptation might prove too strong for some to resist.

Wind farms have become a relatively common sight around the globe, and governments are generally keen to encourage their construction, despite the local protests that can occur (the environmental impact will be considered more closely in Chapter 9). The basic unit is a tower with a large propeller on top, which uses wind to drive a turbine. In order to maximize the strength of the wind, these towers can be as much as several hundred feet high. They need no fuel to run them, and can work either in large conglomerations – one installation in Altamont, California contains over 900 – or individually, to power single homes or farms in remote areas. In some countries wind power provides a significant amount of the electricity: 19 per cent in Denmark, for example, and a substantial chunk in several other European countries. The northern German province of Schleswig-Holstein derives an impressive 36 per cent of its power from wind farms, with Germany being the world's overall largest producer of energy from this source.

The main problem with wind power is that it is inconstant, which cannot help but affect its efficiency. As a case in point, the Danish network had no power at all for a total of 54 days during 2002. Nevertheless, there is increasing investment in wind power around the world, and there is no denying that it qualifies as a clean fuel.

Wave power, too, has come in for more scrutiny of late, particularly in areas with large tidal bores which can provide the power surge needed to drive the turbines. Again, planned schemes often meet with protests on environmental grounds, and these will be dealt with alongside the objections to solar and wind farms in Chapter 9. Wave power is wind-driven, so the same problems of inconstancy and intermittency can arise as with wind power on its own, but it is a source that might well be exploited more in future, being impeccably clean.

Biomass (waste) has also been a subject of development. The principle behind it is that waste is burned to produce steam which then can drive a power station. The planet has no shortage of waste, but the drawback is that when burned it emits carbon, so lacks the clean credentials of other renewables.

The Copenhagen Consensus

Bjorn Lomborg is a highly controversial voice in this debate, and the 'Copenhagen Consensus' of economists that he has been instrumental in founding has given greater weight to his arguments. The Consensus (echoing the 'Washington Consensus', the range of policies which underpins the market fundamentalist system) favours a range of fixes for the global warming problem, most of them fairly low-level in technological terms of reference, thus escaping the riskiness attached to such schemes as orbiting mirrors. There would be little argument that their proposals would be beneficial were they to be undertaken, especially in developing countries beset by a host of health and environmental problems – Africa generally to the fore in such cases. No one is likely to object to schemes for reducing the incidence of malaria, for example, nor to improving the water supply and sanitation systems in those countries either. The problem is whether those schemes would have any determinable effect on reducing our carbon footprint. The Consensus's argument is based on the premise that the situation regarding global warming is nowhere near as serious as the warmers claim, and that the tipping points the latter see looming up just ahead are in fact far away in the future, allowing us to take a gradualist approach to the problem.

Lomborg offers a long list of such projects for our consideration, outlining the technology each involves and providing detailed costings (too detailed for George Monbiot, who remains suspicious of Lomborg on this score, arguing that more guesswork is involved in such calculations than is being admitted[17]). Identifying thirteen key areas in which his smart strategies could be implemented – bringing diseases such as malaria and HIV/AIDS under control, for example, cleaning up the world's water supplies, or instituting research and development programmes into low-carbon energy – he argues that the total cost would be $52 billion per year, as opposed to the $180 billion bill that following Kyoto and other 'feel-good' schemes would run up. An even more persuasive argument from Lomborg's perspective is that his smart strategies would be not just cheaper, but also much more effective – more lives saved, and so on. In no case is the technology complex, nor the organization of a daunting kind to set up; such things are eminently doable if the political will exists. Lomborg's advocacy is powerful and clearly has the greater public good in mind, so he escapes the criticism of self-interest that can be levelled at so many of the sceptical camp:

> [G]lobal warming will probably slightly increase malaria, but CO_2 reductions will do little good compared to direct investments in fighting malaria, which will do 20,000 times more good. And yes, food production will decrease in some places, but if our goal is to fight malnutrition, targeted policies can do more than 5,000 times better by investing directly in hunger prevention.[18]

It is noted that the risk of malaria is predicted to increase anyway, leaving global warming out of account, because of social factors, Lomborg also emphasising, fairly enough, that more risk does not necessarily translate into greater incidence of the disease.

Lomborg makes many striking observations, pointing out, for example, that global warming will mean fewer deaths from hypothermia, which in Northern Europe can be a significant feature of very cold winters. This is something we should welcome, especially since there will be fewer deaths resulting from the increased heat than there would have been from the cold. There is no doubt that Lomborg presents a beguiling argument on the issue, and the statistics can sound quite impressive; per million of population we can expect 1,379 deaths from cold, but only 248 from heat. Thus he can conclude that: 'It seems reasonable from the data that, within reasonable limits, global warming might actually be good for death rates.'[19]

Sadly, one would also have to say that while resource may be forthcoming from rich countries if we do reach unmistakably critical tipping points where positive feedback threatens to overwhelm us unless some action is taken, it is less likely to be so to carry out schemes of the kind that Lomborg and his Consensus colleagues are recommending – all the more so if the global warming threat is perceived to be a distant problem that we can work towards correcting gradually. Gradualism rarely inspires the political class, which needs something much more dramatic to spur it into taking decisive action. Granted, that is an indictment of the rich countries, but political realpolitik suggests that, unless their own survival is obviously at risk, then the required resource probably will not emerge – at least not on the scale that Lomborg and his colleagues are requesting. The humanitarian benefits are obvious, but whether the system will act on that basis alone is more dubious. And as Lomborg presents it, the humanitarian case does seem to dominate. One can agree wholeheartedly with the schemes outlined, therefore (who would be against eradicating malaria, AIDS, or contaminated water sources?), without also conceding that this is the most productive way of dealing with global warming as the current geopolitical order is constituted. It can seem as if Lomborg is fighting a different battle entirely. Stern certainly thinks so, arguing that dealing with what he calls an 'externality' is not the same thing as deciding what to spend public money on; for Stern, the one does not balance out the other.

Living in Denial

There is, of course, another way of living with our carbon footprint, and that is the one favoured by the deniers: do nothing at all. Deniers invite us to ignore the problem, in fact to regard it as a non-problem – or even a blessing in disguise. Hence Nigel Lawson's casual aside in an interview about his book on global warming: 'I think that the ordinary bloke has an instinctive sense that it wouldn't be too bad if the weather warmed up.'[20] That can be a seductive argument, especially if doubt can be planted in the public mind about the reliability of the computer modelling that provides the bulk of the warmers' case – and we know that the latter activity is by no means foolproof. Most people find it hard to think past the lives of their children or grandchildren, and there can be an unreal quality, as Lomborg has noted, to wondering about the state of the globe generations or centuries ahead – never mind millennia. Interest can begin to drift if we do not have a personal stake in the issue. This is a consideration that the warmers do not always bear in mind when issuing their warnings; apocalyptics can be fascinating, but they do not necessarily touch us in terms of our daily lives, often having the air of fiction rather than fact.

The exorbitant cost of anti-climate change measures is another powerful argument with the general public; the sums involved can be colossal enough to induce a sense of unreality, plus an understandable fear that they would probably lead to massive rises in tax rates – never a good selling point for any scheme, particularly one going on seemingly indefinitely into the future (the image of Sisyphus and his labours can come to mind). From the denier perspective, living with our carbon footprint is a relatively minor issue – like living with inflation, for example, annoying but not life-threatening. Lawson's 'wait and see' approach is a variant on this line of argument; do nothing until events force you to, and then only what is absolutely necessary to cope with the immediate situation. Again, one can see the appeal of this to a confused public, more concerned about getting on with their daily lives than events in an uncertain future. The psychology is all too understandable, particularly when confronted with something as dense as the *Stern Review*.

Lawson's bluff manner and anti-scientific bias can be very irritating to a card-carrying warmer, but he is a very representative figure none the less, and one suspects that he speaks for quite a large constituency – the man and woman on the street, we might say – who simply cannot understand what all the fuss is about. At such points the divide between the scientists and the general population can seem very wide – and it is a dangerous divide that must be bridged if anything substantial at all is to be done about global warming while there is still time for it to be effective. Windows of opportunity close eventually. But there are also dangers to be considered if circumstances ever do force us to respond with a substantial programme, or programmes, of action. I go on to ponder these dangers in the next section, where I will be running through some of the worst-case scenarios that could occur if we are persuaded that it is necessary to move past the 'wait and see' position favoured by the deniers.

Part III

The Consequences

7
Worst-Case Scenarios: Economic

Having outlined both problems and solutions, it is time to consider some of the worst-case scenarios that could result from unintended consequences of radical action over global warming: economic, socio-political, environmental, and technological. If globalization were to be severely curtailed, for example, what would be the likely effect on Third World economies? Fairly disastrous, one would be inclined to predict. Those economies are always going to be more vulnerable to shocks than their counterparts in the West, having far less in the way of reserves to fall back upon when downturns occur – and the curtailment of globalization would be equivalent to a permanent downturn for such nations, leaving them in an extremely perilous position. One would imagine that one of the likeliest side effects of such a downturn would be mass migration in search of an improved quality of life; or, to put it more dramatically, to escape from imminent starvation – yet another 'tragic choice' to be faced. This is a topic we shall be following up in more detail in Chapter 8. Before then, short case studies will be drawn from appropriate African countries, such as Egypt and Kenya, to demonstrate the current dependence on globalization in terms of their gross domestic product (GDP), and to speculate on what significant cutbacks could mean for economic and socio-political life – the two can be hardly be disentangled – in those nations.

Another fear is that as climate change causes natural resources – for example, water and fuel supplies, or arable farmland – to dwindle substantially, nations could be prompted to go to war over these in a desperate bid to avert economic meltdown. Such tragic choices just keep multiplying as we survey the troubled landscape of global warming, and we should not assume that the West will be exempt from such problems either.

Tesco Politics and the Developing World

Developing countries rely very heavily on their export trade, and that generally leads to the use of air transport – particularly when it involves perishable foodstuffs. African and South American countries are a ready source of fruit and vegetables out of season, and the Western supermarket system and its consumers have taken to this notion in a big way. The concept of seasonal food has become largely meaningless in the West in recent years, and there is little in the way of fruit and vegetables, whether plain or exotic, that is not available on a year-round basis from one's local supermarket. Meeting this demand has become an important element in the economy of the exporting countries, even if they find themselves being squeezed quite ruthlessly on price by the larger supermarket chains, for whom their shareholders' dividends are invariably at the forefront of their concerns in all such dealings.

The production of fruit and vegetables for Western consumption is problematical in other ways than the mean profit margins that are as a rule offered by the big supermarket chains. Crops require water and that can be at a premium in developing countries. As one commentator has observed, '70% of all fresh water is used for agriculture, so when you buy imported food, you are buying another country's water allocation. Each Kenyan green bean stem is equivalent to four litres of water from a certified "water-stressed" country.'[1] It becomes clear that it is not only carbon emissions we should be feeling guilty about in the West with regard to our conspicuous consumption. The position with meat is even worse, with 1 kilogram of beef requiring anywhere between 100 and 1,000 times as much water in its production as an equivalent amount of wheat would. (As other environmental campaigners have pointed out, meat production also involves an exorbitant amount of grain in the form of animal feed: a powerful argument for a lower-meat diet than the West favours at present.)

It is figures like these that have led some commentators, such as British food policy expert Tim Lang, to make a plea for us to re-examine our diet and to try to wean ourselves away from the driving force behind this, the 'politics of Tesco'.[2] It is an important point, and one that calls for much more public debate than it is currently receiving (every Western country having its equivalent supermarket chain which sets the tone for food marketing internally), but it still leaves unclear what happens to the Kenyas of this world when their export trade fades away because of a shift of priorities in the West. Reduced stress on one's water supplies, or less foreign exchange: not quite a tragic choice perhaps, but not exactly the easiest of trade-offs to resolve either. Nor is it one likely to become any easier as persistent drought takes hold of so much of the developing world.

Even if trade were maintained with countries like Kenya, that would not be the end of the problem. Global warming has other implications for agriculture in tropical regions. As Tim Lang has noted, climate models suggest that 'a one degree rise in temperature can lead to 10% yield reductions in tropical crops,' and that cannot be good news for the developing world, where agricultural exports can be a vital part of the GDP.[3] Such crops might then become viable in higher latitudes, as we are seeing happening with regard to vines, but that merely worsens the plight of the tropical countries, who would not even be able to hope for higher prices for scarce items if they were flourishing elsewhere. The supermarket chains will simply follow wherever the crops happen to go, after all. The consequences of a temperature rise of several degrees hardly bear thinking about.

Globalization and GDP in the Developing World

What does globalization mean in terms of the GDP in African countries such as Kenya or Egypt? Kenya has a very significant horticultural export trade (that is, fruit, vegetables, and flowers), worth $1.1 billion dollars a year. Combine this with the country's other main agricultural exports, tea and coffee, and you have roughly half of its total foreign earnings. As an indication of the size of the horticultural trade, 35 per cent of the roses sold in the EU come from Kenya. Baby vegetables from Kenya are also a very popular item in British supermarkets and in up-market British restaurants. Africa in general provides 14 per cent of horticultural imports into the UK, and although this has caused concern about its carbon footprint, a recent study by the Department for Environment, Food and Rural Affairs (DEFRA) has pointed out that 90 per cent of this comes over by ship rather than by air freight. Since ship transport has a lower carbon emission per tonne of freight than air does, DEFRA concludes this is an acceptable practice (although that is not to say that shipping emissions are negligible, as we have seen; it depends on the volume of traffic). They also emphasize that African exports in this area constitute only a small portion of food and related products' transport costs in the UK anyway, so they are fairly sanguine about the situation.

But there is still a problem to be noted in that, as the DEFRA report concedes, '[h]igh value, perishable products such as leguminous vegetables and cut flowers tend to be air-freighted,' and it is in these that Kenya specializes.[4] DEFRA nevertheless feels that the relatively small scale of this trade makes the air freight bearable in this instance, and points out that such trade is commercially very valuable to a country like Kenya, which would feel the pinch if it were curtailed significantly. It also refers to a study conducted by Cranfield University which found that roses imported from Holland into the UK involved higher carbon emissions overall. As we are becoming increasingly aware, the entire area of carbon emission solutions is full of ironies such as this, and the obvious answer to particular problems does not always prove to be the right one. Even air-freighted vegetables are not necessarily all that much more carbon-intensive than their greenhouse-grown equivalents would be in the UK – only about 15 per cent as Fred Pearce points out, so he now feels that he can 'buy Kenyan [green] beans with an easy conscience', air miles and all, despite the protestations of his ultra-green friends.[5]

Egypt's main exports are oil, cotton, and agricultural goods (Egyptian new potatoes are a common item in British supermarkets, for example), and its export trade was worth $13.8 billion in 2005, representing a substantial 31.4 per cent increase since 2003. While moving towards market liberalization, the country is running quite a substantial trade deficit, with imports in the same year amounting to $24.2 billion. The West, in other words, is doing very well out of trade liberalization, with a substantial chunk of Egypt's deficit coming from the import of Western products, or at least Western brands. That is another danger of globalization as it is currently practised, that the local economy will be swamped by incoming goods once trade and financial controls are relaxed, and thereafter find itself quickly sliding into debt (debt relief is already a source of much contention in world politics). Many of these Western brands – clothing, for example – will have been manufactured at low cost in the developing world, only to be exported back there at much higher cost: yet another way in which the latter loses out in the globalization merry-go-round. This is not so much an argument against globalization, however, as its current style, which demands a totally open, control-free market, and makes it difficult for countries which do not comply. The system can be altered to be more equitable, with experts like Joseph Stiglitz providing various proposals as to how this could be done (fairer prices for raw materials and natural resources, helping local economies to develop, more debt relief, and so on). There needs to be less ideology and more flexibility in the global market.

If the West did opt for a retreat from globalization, however, that would give countries like Egypt no opportunity to overcome their trade deficit, condemning them to permanently underdeveloped status with no escape from the poverty trap that would inevitably ensue – and we should note that Egypt is already experiencing food riots from rapidly rising world prices. The workings of globalization are fickle enough for developing countries as it is (the Egyptian cotton industry is in crisis at present owing to plummeting world prices, despite the very highly regarded quality of its product), but were the system to be run down in order to inhibit carbon emissions their situation would become far worse – unless, of course, they were able to make use of their natural resources in the production of renewable power, as Egypt clearly could. The Sahara has been identified as a prime site for the development of solar power, which could then be sold on to Western Europe as an alternative to fossil fuel. We have already seen how the south-western deserts in the USA are being geared up for an expansion of solar power production, and the Sahara offers equally ideal conditions. Egypt could well manage to overturn its trade deficit if it was able to develop this method on a large enough scale – and the Sahara certainly provides that scale.

Other countries in the Sahel region could draw similar benefits from exploiting this renewable resource, as could the adjacent Middle East, where it could in time come to replace oil revenues to at least some extent. There are, however, potentially very damaging environmental costs to be taken into account when converting deserts to the generation of solar power, which we will be considering in more detail in Chapter 9. There is rarely a simple fix in this game; trade-offs always come at a price, and whether it is a price worth paying is not always easy to determine beforehand.

Globalization and its Western Outposts

It is not just the developing world that would experience problems if global trade were to be run down. The West is an elastic concept which takes in societies such as Australia and New Zealand, which in effect can be considered Western outposts. Although these countries have closer links to their Asian neighbours than hitherto, they are still culturally very close to the West – not surprisingly, given their overwhelmingly European-derived population. Trade with Asia may have increased considerably in recent years, but even so both Australia and New Zealand would be hard hit by a cut in their exports to the West. New Zealand in particular has a very small home market, and its trade in lamb, a product well represented in the British supermarket system, is one of its staples; there is even a New Zealand Lamb Day to emphasize its importance to the economy. The country's sheep and beef export industry is worth $5 billion annually, and it is the world's largest exporter of lamb. Lamb is produced extensively in Europe too, and in terms of food miles to transport it from New Zealand, one of the furthest points on the globe that Europe can export from, might seem difficult to defend. Yet to do so apparently represents a saving in emissions, although that might suggest that the industry in Europe is being very inefficient and should be examining its methods with a view to reform.

New Zealand is also an exporter of agricultural products and wine, and one would have to assume that in theory these could be produced at lower emission cost in the West. The same can be said of most of Australia's exports. Whether such fine-tuned calculations would continue to be made if retreat from globalization became the order of the day it is hard to say. An interesting aesthetic consideration comes up in this context. Wine is produced in substantial quantities in the West (the EU has even had 'wine lakes' in the recent past owing to surplus production), but if Australian and New Zealand wine tastes different to its European counterparts, can it be treated as exactly the same product, to be judged equally according to its carbon footprint? Australia is also building up a reputation for its olive oil, but does that mean we have no real need of it in the West given our own numerous local sources of the product? To a connoisseur, neither wine nor olive oil is simply a generic product (like petrol, say), but something with local characteristics which affect the taste very markedly – and one can develop distinct preferences in this respect. The food miles issue can become somewhat complicated if products are not really identical and there is an important aesthetic dimension to be taken into account (and wine comes from all over the globe these days, the best of it with its own local character to distinguish it from its competitors). How meaningful that dimension would be in desperate circumstances is, I concede, another matter, but there is no doubt that something would be lost if there were to be any severe geographical restrictions on our diet.

The Global Resource Squeeze

International Alert's *A Climate of Conflict* paints a very depressing picture of a world of declining natural resources and the political problems that scarcity is already beginning to create for many societies – 'overshoot' at its most critical. Given that extended periods of drought for various areas of the globe are part of most projections as to the effects of global warming (often very extended), this is a disturbing prospect indeed to contemplate. Drought has been a fact of life in the Sahel and Australia for quite some time now, and we can only assume it is a condition which will become ever more common as we move through Lynas's +1–6°C degree scale (although in some cases it will lead to more rainfall, if not always where that will be most useful). We know that drought has been a major contributing factor in the collapse of sophisticated civilizations in the past, and there has to be a real fear that it could have the same effect again. Without reliable water sources societies are placed under intolerable stress, and if the problem is not remedied then something most likely has to give in their socio-political structure. That is a state of affairs often leading to conflict, both internally and externally.

In Australia in particular, we can find an example of a Western-style society having to face up to the unimaginable in the wake of a drought that has left large areas of the country without significant rainfall for several years – the unimaginable being its inability to sustain the continued existence of its major centres of population, such as Sydney or Melbourne. The problem goes hand in hand with the irony of Australia's largely unpopulated north-east corner, with its tropical climate and extensive rain forests, being predicted to receive increasing rainfall as climate change progresses – although that does raise the question again of the reliability of regional forecasting through climate modelling. Whether this will dictate a shift in population or large-scale engineering to pipe the water from the north-east to the parched south-east where the bulk of the population lives (and suffers from increasingly dangerous drought-induced bush fires reaching right into the city suburbs) is an open question, but the actual viability of Australia as a modern nation state is now an issue for debate. Add another degree or two on to global average temperatures, as seems unavoidable the way things are going, and we may well slide past the tipping point on that, whether or not action is taken to match up population with water supply.

It is not all that difficult to sound quite apocalyptic about Australia's future; again, it seems a case of hoping for the best but fearing that the worst could happen – and at not too distant a point ahead either. Jared Diamond's warning based on the experience of the Maya civilization in central America is worth heeding: 'Lest one be misled into thinking that crashes are a risk only for small peripheral societies in fragile areas, the Maya warn us that crashes can also befall the most advanced and creative societies.'[6] The more complex a society is, then the more vulnerable it becomes to system failure (something that we have learned from chaos theory); and it does not take very much to go wrong to destabilize a complex system. Computerized systems alone should teach us that; there can indeed be chaos when these go down, as when a virus strikes. Drought functions much like that virus.

The traditional West as well may have to get used to bitter disputes over resources such as water, something we have never really had to face in modern times. In Spain, water shortage is already creating political tensions between the region of Catalonia, which barely regards itself as part of Spain at all and has a long tradition of campaigning for independence, and the central government in Madrid. Catalonia is in the midst of a chronic water shortage and has had to appeal to the central government for help in resolving it, with the Catalan regional government head feeling moved to plead that 'Catalonia is also part of Spain' in order to prompt action.[7] Spain's answer to its increasingly severe water problem nationally has been to invest heavily in desalinization plants, which now provide much of its fresh water supply. But as the Spanish Association for the Technological Treatment of Water has pointed out, the operation of each plant leads to the generation of a significant amount of carbon emissions – around a million tonnes a year (and Spain already has 950 such plants, with more planned). Bearing this in mind, critics have suggested that the diversion of rivers to those areas most in need may be a better policy to adopt, but the effect of this is to damage ecosystems with unknown longer-term consequences (a familiar refrain, as we are finding out).

Battle-lines are being drawn over the issue, which promises to have repercussions for some years to come in Spanish politics. For the time being, Catalonia is shipping in water by tankers, but that is not an emission-free solution either, nor necessarily one that can be kept up indefinitely. Apart from the expense and the logistical problems of distributing it, there has to be water available for sale – and at affordable prices.

The chances are that Spain will not be the only Western country to face a resource distribution problem, nor the last to find it turning into a divisive political issue internally. The entire Mediterranean has to be considered at risk when it comes to water supplies, especially since most modelling projections see the European side of the Mediterranean becoming more like the African as temperatures keep edging upwards. This would turn much of the area into little better than desert scrubland (as in Morocco and Algeria, for example), destroying much of its agricultural production in the process. We have already noted how vineyards are gradually creeping northwards; another implication of this trend is that they will be burned up in the south where they currently flourish. The hotter it becomes, then the more irrigation that will be required to keep crops like vines growing healthily; much of the Australian vineyard area is only viable through intensive irrigation already (although whether that practice can be continued given the nation's growing drought problems is clearly an issue of some political importance). But if the water supply is drying up too, then the situation rapidly becomes untenable. There could be a mass resort around the Mediterranean to desalinization plants on the Spanish model, but of course this would merely increase the carbon emissions that were one of the main causes of the problem in the first place – as would shipping in water by tanker on any such potentially massive scale. None of this would exactly be good news for the region's economic life.

Some countries are also already thinking ahead about how best to protect their own resources at the expense of the needs, and one might say desperate needs too, of their neighbours. In the Sundarbans delta in the Bay of Bengal, where, as was noted in Chapter 2, entire islands are already disappearing because of flooding generated by the effect of global warming in the Himalayas, India is, as one report puts it, 'sealing itself off from Bangladesh', by building huge fences and barbed-wire barriers. Not much evidence here of Nigel Lawson's claim that a little warming is an event to be welcomed. Ultimately, the entire edifice is planned to reach the impressive – or obsessive, depending on your perspective – length of 2,500 miles. The objective is to prevent Bangladeshi refugees trying to escape into India from the increasingly catastrophic flooding in their already badly flood-prone country.[8]

No doubt the cynical calculation behind such a project is one we shall probably have to become accustomed to in years to come as well, as the effects of global warming, whether in the form of flood or drought, become ever more evident around the globe. The early signs are that this will not do much for the development of human compassion or our sense of community, never mind our concept of ethics, and that is not exactly a pleasant prospect. If it is to be everyone for themselves, with cosmopolitanism and risk-sharing being kept off the agenda altogether in favour of protecting one's own resources at all costs against one's neighbours, then we face a very bleak future. Policing the barrier will be no mean task either, and one that will almost inevitably lead to violence of some sort; the afflicted will go to desperate lengths if it becomes a question of their actual survival, and there would seem little doubt that the barrier will be the subject of regular assault as the environmental situation deteriorates. Political relations could only deteriorate too under such circumstances, ratcheting up global tensions even further.

The Indian/Bangladeshi barrier is a physical one, but the West would be putting up barriers at least as forbidding with the developing world if it cut back radically on its exports from this quarter: one of the 'virtual fences' that Naomi Klein sees being erected all round the world to protect Western economic interests at the expense of vulnerable local populations.[9] It would not take much to engender a worst-case economic scenario for nations like Kenya and Egypt – and a host of others in Africa, South America, and Asia as well. Countries like that would be facing collapse, their lifelines cut. Without substantial trade with the West they would be unable to sustain themselves economically, with all the implications that would have for their survival as political entities. A world with a string of failing nations in it would not be a particularly safe place, and economic collapse would trigger a series of socio-political worst-case scenarios, as I will now go to consider in more detail.

8
Worst-Case Scenarios: Socio-Political

If mass tourism were to disappear as a global phenomenon, what would be the social and political implications? Fear of the (cultural) other is engrained enough in humanity as it is, and would be unlikely to decline if contact between various races and cultures became ever more restricted. Mass tourism brings many problems in its wake (pollution, pressure on scarce resources like water), but it nevertheless promotes contact between societies, and that is always valuable. If such contact is admittedly often of a somewhat artificial kind, centring on conspicuous consumption by the one side and the servicing of that on the other, it is enough to keep even the most prejudiced aware of the existence of cultural difference and cultural diversity – perhaps to come to respect these, even if somewhat grudgingly. Cut this back and the likeliest consequence is an increase in racist attitudes, which are present just under the surface in most Western societies anyway – and elsewhere too. This would simply play into the hands of the more reactionary forces in such societies, and thus threaten the liberal ethos that has transformed Western culture from the Enlightenment onwards. A drift back into political authoritarianism could be a very unwelcome unintended consequence in this respect, and one that unquestionably deserves to be resisted; but certain circumstances could make it a much more formidable opponent than it is at the moment. Crisis often generates a rise in authoritarian and totalitarian attitudes – out of desperation as much as anything else.

If Third World economies proceeded to collapse because of radical action taken to curb global warming, a very likely effect would be mass emigration by their populations. This is a traditional human response to environmental problems, although it would be far more complicated to put into practice now than in earlier eras, when populations were very much lower and territorial consciousness was too. Nomadic behaviour is an anomaly in an advanced capitalist culture such as the one the majority of the world now inhabits, as the Romany community would be the first to attest, given their well-documented problems in maintaining their traditional lifestyle throughout Western Europe. The West already has a sizeable immigrant community from the Third World in its midst, and claims that this is overstretching its resources – welfare provision, housing, and employment being the most usually cited – as well as creating considerable social tensions. Riots and gang warfare between host and ethnic minority communities are a regular occurrence in Western European life these days, France and the UK being particularly prone, and one that is causing increasing concern to the authorities. Immigration has turned into a high-profile political issue throughout Western Europe, and various new methods of control are being introduced – even against other EU member states (from the old Eastern bloc, for example, which is gradually, and in some cases only with considerable reluctance on the part of the older community members, being absorbed as latecomers into the EU family). What would be the social and political fallout from a really massive increase in immigration, most of it probably unofficial, and how might this alter the balance of global power?

The Tourism Equation

Tourism is currently one of the world's biggest industries, estimated to be worth around $8 trillion in 2008 according to the World Travel and Tourism Council (WTTC), who forecast that this figure will rise to $15 trillion by 2018.[1] WTTC claims that tourism generates almost 10 per cent of GDP globally. The industry has been growing exponentially in recent decades, thanks in large measure to the advent of cheap air travel. Most of the globe is now covered by budget airlines and airline fares are now a fraction in real terms of what they used to be; often, bizarrely enough given the discrepancy in their respective carbon emissions, far cheaper than rail tickets for the same journey (the UK has a particularly bad record on this score, that betrays yet another lack of coordinated planning in dealing with climate change). Weekend breaks in exotic places have become part of the everyday lifestyle of much of the population in the West as a result. The budget airlines regularly complain about being singled out for criticism as regards carbon emissions, claiming that they are responsible for only a very small percentage of the overall global total (their fraction of the 1.6 per cent that air transport contributes). That may be true, but it is a contribution which has been steadily increasing and it can hardly avoid being taken into account in any debate over the topic – especially given the enthusiastic official commitment to airport expansion throughout the West (not to mention the particular potency at high altitudes of airline emissions). London Heathrow, as a high-profile case in point, has a vast new terminal and is still campaigning for a new runway, despite the environmental disruption this would cause and the spirited opposition it has generated from environmental campaigners.

What might happen if budget airlines were taxed to the point where there was a dramatic fall in custom? Massive increases in fuel tax that would lead to very much higher fares have been touted by environmentalists as one way of bringing about this state of affairs, and it has to be seen as a distinct possibility if public pressure grows for politicians to do something positive about carbon emissions rather than relying, as most do at present, on voluntary controls. Whether politicians will feel able to go to a level that really does cut custom, as opposed to creating a temporary dip (the usual outcome when other leisure-time products such as alcohol and cigarettes are taxed more), will no doubt depend on how serious the situation has become in the interim – and also whether they feel they can trust the public mood. The electorate can be very fickle in such cases, apparently wanting something to be done, and saying so in polls and surveys, but not willing to face the consequences in their personal lives, such as having to forego holidays in exotic places. They can be equally unreliable when it comes to oil, accepting that we should use less of it but complaining bitterly if the price goes up and they actually have to countenance a cut in their own consumption. Direct action over petrol costs is not unknown, and this is an area where governments tend to tread very warily; motorists are a powerful lobby.

There are various countries around the world which are heavily dependent on tourism for their economic survival. Africa has several examples, such as Gambia, where carefully patrolled tourist areas keep the country's acute poverty (it has one of the lowest average incomes in Africa) at arm's length while visitors indulge themselves in the traditional holiday joys of sunshine and beach. The Caribbean, too, has several examples of the dedicated resort phenomenon, and while one can be critical of this ('tourism apartheid' being one of the descriptions) there is no denying that they do contribute to the local economy. Not as much as they might perhaps, but that is another issue; as the Secretary-General of the United Nations World Trade Organization (UNWTO), Francesco Frangialli, has insisted, 'tourism can play a major role in improving the standard of living of people and help them lift themselves above the poverty threshold.'[2] Tourism is also a big factor throughout South America, where it can form a very significant percentage of the GDP of many nations; in Peru, for example, it constitutes 7.5 per cent, with sites like Machu Picchu being tourist magnets.[3]

Going back to Gambia, tourism is a recently developed industry there, generally considered to have begun in a organized sense only with the trip set up for 300 Scandinavian tourists by the tour operator Vingresor in winter, 1965. Tourism very much plays to the country's climate strengths, and Gambia has turned into a popular 'winter sun' destination, particularly for Northern Europeans. The industry is now trying to diversify into ecotourism, however, which taps into a different market and has the added advantage of involving local communities in the hinterland rather than just those around the beach resort hotels. (On this last point, it is more than somewhat ironic that one report on the tourist industry in the country has approvingly noted that the government's new green tourism policy 'should help increase scheduled air flights to Gambia':[4] one step forward, two steps back on the carbon emission front, one would be inclined to say.) At present, tourism represents 13 per cent of Gambia's GDP, and provides 19 per cent of jobs in the private sector.

The extent to which tourist revenue filters down through the population at large is, as always in the developing world, difficult to determine with any great degree of accuracy. Studies by organizations such as the UK's Overseas Development Institute suggest that much of it in this instance goes to the tour operators and the airlines instead.[5] That is a situation which calls for change, but remove tourism from the country's revenues altogether and these would no doubt decline very considerably, with predictably dire consequences for those at the lower end of the social scale, who have little margin for any drop at all in their income or whatever meagre government support they may receive. Even tragic choices have a lesser to their evils.

Although I have said that tourism plays to Gambia's climate strengths, whether these will continue to be perceived as strengths if temperatures continue to climb ever upwards is another matter. There has been speculation that we may even start to find cooler regions of the planet more attractive as holiday destinations if the traditional ones, not to mention our own home countries, become progressively hotter as global warming advances: Greenland lake district anyone? Tropical regions will, of course, be the first to feel the impact of any such change, with even Northern Europeans seeking a sunshine fix likely to find the heat becoming intolerable – especially if Europe is going through summers like 2003, or worse, on a regular basis. The novelist Thomas Hardy once observed that he could envisage a time,

> when the chastened sublimity of a moor, a sea, or a mountain will be all of nature that is absolutely in keeping with the moods of the more thinking among mankind. And ultimately, to the commonest tourist, spots like Iceland may become what the vineyards and myrtle-gardens of South Europe are to him now[.][6]

Our current bout of climate change might well make this a reality, although not for the reasons that Hardy imagined. We may soon be making our way on holiday to Iceland and points north in general to escape the heat. By then the vineyards and myrtle-gardens of Southern Europe may lie scorched and barren, no more than a distant memory – never mind the tropical areas where we currently go. Winter sun may well come to seem a perverse notion when vineyards are flourishing in areas like Finland – and who knows how much further north they might go?

Neither is this dependence on tourism restricted to the Third World. Many European nations would face quite severe economic problems were the mass tourist trade to decline markedly (or decamp to points north instead). One has only to consider Greece, where in 2008 20.9 per cent of the population (1 in 4.8 jobs) worked in the tourist industry or related activities. This is expected to rise to 21.9 per cent (1 in 4.6 jobs) by 2018. Tourism constitutes 17.2 per cent of the country's GDP at present, and is predicted to increase by an average of 3.9 per cent per annum over the next 10 years.[7] Greek economic life very largely revolves around tourism, but even more developed economies would feel the pinch if tourism were to decline. The likelihood is that no European country would escape a very significant fall in revenue, especially if transatlantic flights were to drop sharply in volume. In France, for example, the tourist industry is estimated to be a $307 billion business (13.1 per cent of jobs).[8] The UK had 32.6 million overseas visitors in 2007, who collectively spent £16 billion, helping to maintain two million jobs overall.[9] Given figures like these, it is clear that the fallout from any really notable decline in visitor numbers could be very damaging to the GDP of such countries. Unless there are significant advances in producing clean fuels for transport in general, there are bound to be problems ahead in the tourist industry. For all their recent success there are fears that budget airlines could face difficulties as fuel prices soar for market reasons (price hikes by members of the Organization of Petroleum Exporting Countries (OPEC), for example), and that would have a substantial effect on tourism throughout the West as well.

A switch to more upmarket tourism – as ecotourism tends to be – might resolve the issue to some degree, keeping the industry functioning, but whether this could yield the same income all round as mass tourism does is more questionable. The likelihood is that such a market would be far less labour-intensive, which would not be good news in employment terms (poorly paid and menial though much of that employment generally is).

The Racism Factor

In principle the West is now multicultural and racial discrimination is illegal. Anti-racism laws are in place in most Western countries, and real progress has been made on that score in recent decades. The populations of European nations are ethnically more diverse than they ever have been, and places like London and Paris are now truly world cities, with most of the globe's societies being represented there to at least some extent, injecting their own culture and cuisines into the metropolitan mix. Many of us find this an entirely welcome phenomenon, appreciating the vibrancy and variety that multiculturalism brings to our lives.

Yet whatever the official line, multiculturalism has been a source of stress and tension in most Western societies, and it does not take much for immigrant communities to be scapegoated when problems arise. There only has to be a slight economic downturn for anti-immigrant sentiments to surface, and for questions to be raised about the qualifications required to settle in areas like Western Europe. In fact, it is becoming increasingly difficult to meet the criteria that are being set for immigration into the longer-established EU countries, which generally demand a high level of skills or personal wealth in order to gain entry from outside. One category that is particularly unpopular with officialdom is economic refugees: those who are fleeing their own countries because of poverty, often to the extent of starvation – from drought-ridden Africa, for example. Economic refugees are exactly what climate change promises to deliver in ever-increasing numbers in the near future, however, and that could turn out to be a worst-case scenario for everyone involved.

The assumption has to be that racist tensions would be heightened considerably by a greater influx of immigrants – many of whom would be there illegally, just to add to the atmosphere of ill will. Already, as we have observed, there are problems to report. In the UK, anti-immigration arguments have been put forward by the main political parties, and fringe organizations such as the British National Party (BNP) have exploited race to some effect in local elections, where the issue seems to resonate particularly strongly with voters. The popular press tends to present immigrants as a drain on the country's resources, particularly in terms of their use of health and welfare services, the claim often being made that they are taking out far more from the country's economy than they are putting in. Stories about the strain on the National Health Service from dealing with immigrant communities are fairly common, and help to stoke up a feeling of resentment against them amongst the general public that hardly helps community relations. One could blame the media, but there has to be something for them to work on – and they do so with no little dedication.

Mirroring this general sense of suspicion about immigrants, government policy is very much directed towards selectivity in immigration, favouring those applicants possessing skills that are officially recorded as in short supply – with language skills a part of the package – over the claims of economic refugees. More menial occupations rarely get much of a look in, although economic refugees are the most likely to fill such posts, generally coming from the poorer, less educated sections of their home society. Even if it is difficult to find local inhabitants to perform such tasks, particularly low-paid cleaning and catering jobs in cities like London, there can still be resentment about the fact of immigrants having any employment at all. If that can happen now, just imagine how much worse it could become if the ranks of economic migrants continued to swell year after year as conditions worsened elsewhere on the globe. Western Europe is fairly densely populated as it is (Holland and England being prime examples), and all countries must have something like a saturation point when its systems just no longer cope very well.

If overseas tourism declines sharply then there will be less contact between Europeans and other races in situations where the latter are more likely to be accorded a measure of respect, being on their home territory. In such cases at least a basic degree of civility is usually forthcoming from the Western tourist, even the more prejudiced examples of such. Travel does not always broaden the mind as much as one might hope, but it generally does so to some extent from seeing how others live and organize their society, and that opportunity for broadening could well be lost for a significant section of the population if mass tourism were to be scaled down dramatically. Other races will only become more mysterious if we encounter them less, and will tend to be seen as a threat when they migrate to the West (whether legally or illegally), their motives constantly under suspicion, however unjustly – 'taking our jobs away', and so on. All of this will provide exactly the kind of context that the reactionary forces in Western society would welcome, and they will do their utmost to exploit that suspicion for political gain. With a bit of deft manipulation latent prejudice could easily be brought to the surface.

The cause of race relations could well be put back several generations if this worst-case scenario were to occur, and it would not take much for that to become reality. Nor would it just be the cause of race relations that was put back; the liberal societies we have constructed in the West with their strong concept of human rights would also be under threat from any upsurge in reactionary politics. We have seen before how effectively fascism can work its way into the political mainstream if economic conditions deteriorate badly, and we could run that risk again if Western European societies were put under really severe population and resource stress.

It must be emphasized that this is not meant to be an argument against immigration, which is generally to the advantage of the host country, no matter what the reactionary forces may say (I am talking about experiences up to the present only, of course, involving manageable numbers). The arguments of those forces should be challenged across the board; racial prejudice is simply wrong and should not be tolerated. But immigration is not the same as forced migration; the former is a proper choice, the latter a tragic choice that should not have been necessary. We should be working to create a situation where as few tragic choices as possible need to be made, especially if the exercise of those tragic choices brings out the worst in human nature. Sadly, that is something that climate change has a considerable capacity to do. There would be a real test of our social conscience if the population of the West were to soar, especially when resources did come to be dangerously stretched – and resources are clearly not infinite. Whether that social conscience would survive such an event really needs to be considered, but International Alert's warnings about resource wars could have a resonance well outside the impoverished developing world.

All Change?

History is full of mass migration by peoples, and even if this is more difficult to initiate nowadays than in the past it will still occur if the environmental situation becomes desperate enough. Given that the poorest nations constitute a majority of the world's population, the probable scale of such migration in our own time would be alarming to contemplate. If we recall International Alert's estimate that the lifestyle of 3.9 billion of the world's population is already under threat from the effects of climate change, then we are forced to realize just how many candidates there might be for mass migration. It is unlikely that such disruption could take place peacefully, and the potential for conflict would be very high indeed if there was a huge shift of population from south to north, east to west, all of it fleeing from a combination of socio-political and economic breakdown brought on by environmental degradation and a retreat from globalization. There is no real precedent for this in the modern world in terms of the likely numbers that would be involved, and it would be difficult to predict what the world political map would be like after any prolonged episode of migration involving even just a substantial minority of the 3.9 billion pool. Large parts of the world could lie abandoned, with the population being forced into ever smaller and more densely inhabited areas (mostly in the north), with social tensions increasing exponentially in consequence. There would be a full-scale agricultural crisis to accompany this, as so much arable land was lost to the effects of climate change. If the land that was left was also being used to produce biofuel, then so much the worse all round.

There have been speculations that the world population might begin to fall, since in many of the developed countries the birth-rate is beginning to dip below the level required to replace the dying. Fred Pearce has even wondered whether this might lead to certain countries having to woo migrants in order to prevent significant labour shortages occurring.[10] But that assumes a gradual readjustment, and not the kind of situation that is being predicated here (and although it would be helpful for the global population to fall, there is no agreement amongst demographers over that issue either).

Economic refugees are a common enough phenomenon in Western countries as it is, but if a majority of the globe's inhabitants are turned into truly destitute examples of this then our system could be heading for a breakdown as well. Neither socially nor politically are we equipped to cope with such a situation and we should be doing our utmost to prevent it occurring – certainly on the scale that I am suggesting is possible if we were to take drastic action against global warming. One of our main methods of attempting to hold warming at bay will be through the application of cutting-edge technology, but, fairly predictably, that brings its own set of problems in its wake, as the next chapter will show.

9
Worst-Case Scenarios: Technological and Environmental

Environmental engineering has many supporters within the scientific community, who relish the challenge it poses to their knowledge, theories, and ingenuity, as well as the chance to raise their public profile: 'scientists come to humanity's rescue', as the media would no doubt report any success story in this line. But how safe is such engineering, and how predictable is it in terms of its application? The answer is that no one really knows until it is tried, and after that it is too late to do anything very much about it if it goes drastically wrong – always a distinct possibility with such speculative procedures as altering the amount of solar radiation the Earth receives. Measures taken to reduce surface temperatures on the planet could generate adverse weather systems affecting crops, or even achieve the exact opposite effect to that being sought – increased rather than decreased warming. We are back with the problem that we do not have comprehensive knowledge of how all the parts of the system interact, and as the 'butterfly effect' in chaos theory tells us, even small changes to a complex system can have catastrophic effects on it as a whole, rapidly amplifying out of our ability to control (a butterfly's wings beating in one place creating a tropical storm several thousand miles away, as the famous example has it). Any attempt to manipulate the weather has the capacity to backfire quite spectacularly (ecocide on the grandest of scales), and although the technology already exists to try it out, as well as the computer modelling systems to make forecasts as to the likely outcomes, caution so far has won the day and nothing really large-scale has yet been put to the test. Nevertheless, schemes continue to be put forward for our consideration.

Solar and wind farms and their impact on the environment (including human health) deserve to be scrutinized closely, too. What might be the unintended consequences of colonizing vast tracts of wilderness for such projects? Such schemes are under way, as in the south-western states of the USA, where the solar farm system is already well advanced as a method of energy production, with plans for considerable expansion in the near future. And what about the aesthetic dimension of wilderness: how much consideration should we give to this?

Biofuel has been much in the news of late, and is attracting substantial interest from several governments; but there are environmental dangers to the spread of its production as well, particularly in the developing world, although an awareness of these is now growing. Thus the Ugandan government, under pressure from conservation groups, has (2007) drawn back from a scheme to convert the biodiversity-rich Mabira forest near Victoria Falls into a biofuel production area.[1] Whether other Third World governments will feel able to place the preservation of biodiversity over biofuel profit margins remains to be seen, but one would probably not want to bet on that either – even in the short term (and indeed the Kenyan government is trying to push through a similar scheme in the Tana River wetlands in 2008, despite concerted opposition[2]). The pressure to produce more biofuel as we reach peak oil will become more and more intense, and it will take a strong-minded government to resist its profit lure. Whether biofuel really is the answer to the fossil fuel crisis is something I will go on to consider later.

It's All Done with Mirrors

Let us say that circumstances become bad enough for us to decide to go ahead with some of the schemes designed to reduce the amount of solar radiation our planet receives; what could go wrong? To reiterate the principle behind the idea: a significant quantity of incoming solar radiation will be prevented from reaching the Earth and becoming trapped in a greenhouse gas-saturated atmosphere. In the process, a balance would be achieved such that temperatures would not rise despite the doubling of the amount of greenhouse gases in the atmosphere since the beginnings of industrialization two centuries or so ago. The basic problem is that tinkering with parts of the system can alter its workings elsewhere for the worse – a recurrent theme in terms of response to the problem of geoengineering.

One recent modelling exercise, carried out by scientists at the University of Bristol, suggests that a 'sunshade geoengineered world' would not respond in the same way that a pre-industrial world did to similar global annual temperatures, predicting a drop in global rainfall of 5 per cent, for example, which would hardly be helpful to an already drought-prone system.[3] The study involved simulations of Earth's climate in the pre-industrial age, with carbon dioxide levels at four times that value, and then the latter with a 4 per cent reduction in solar radiation thanks to the use of sunshade geoengineering. The team's findings were that the Earth did not return to its pre-industrial climate, but that there was a differential reaction to the sunshading: the tropics became 1.5°C colder than then and the higher latitudes 1.5°C warmer. One consequence of this would be less sea ice, and therefore a knock-on effect in terms of the ocean ecosystem. Modelling always has its drawbacks, but this is the most detailed study yet undertaken as to the effects of successful sunshading on the environment, as opposed to just striving to achieve a reduction in solar radiation on the grounds that this action alone would suffice to save us. It is further evidence, if we needed it, of just how complex the organism is that we are trying to geoengineer; to do so can seem like something of a lottery.

Problems could also arise with the sulphate solution, whereby sulphate particles would be injected into the atmosphere on a large scale to reflect sunlight back into space, in the manner that volcanic eruptions are known to do. Recent modelling of this idea has identified some potentially very unwanted side effects. The sulphate particles 'could destroy between 22 and 76 per cent of the wintertime ozone layer over the Arctic', the modeller, Simone Tilmes of the National Center for Atmospheric Research in Colorado, warns, as well as holding up the recovery of the massive hole that has been created by chlorofluorocarbons (now largely banned in the West) over the Antarctic by anything up to 70 years.[4] Since holes in the ozone layer lead to increased incidence of skin cancers from exposure to ultraviolet rays, this raises the prospect of yet another agonizing choice being put before us. If nuclear power means more cases of cancer, do we want to create even more through geoengineering? Other speculations have been that more sulphate particles in the atmosphere could increase the amount of acid rain, which would be another deeply unwanted side effect – they do seem to keep piling up, the more detailed the modelling (although it is only fair to record that there are sceptics regarding the extent of the danger that acid rain poses for us[5]).

Games with the Desert

Albedo enhancement of the Earth's surface (GAEP) was put forward as an alternative to geoengineering in space, but even its deviser, Alvia Gaskill, had to admit that her proposal to cover deserts like the Sahara and Gobi over with white plastic sheeting for at least sixty years had several drawbacks. In her words, covering 'removes the land for other uses for possibly hundreds of years' and 'will kill all plant and animal species in the covered areas'.[6] This is bad enough, but add to it that covering may alter the climate for the worse where it is in operation, and even lead to the desert expanding in areas like the Sahel, and we might wonder whether the side effects render the idea untenable. As described it would be a disaster ecologically, and we might also note that it would put out of commission some of the most appropriate areas in the world for the development of solar power facilities. A Sahara covered in either huge solar power stations or white plastic sheeting begins to sound like yet another tragic choice to be made, ecologically and aesthetically. If the Sahara were also to expand significantly in size, then GAEP would have the potential to project us into a large-scale socio-political disaster as well – and that in an area which is teetering on the brink of it already for environmental reasons: a worst-case scenario all round.

Once again we would have to rely heavily on modelling, and that always involves a considerable element of doubt as to outcomes in the real world. Although some field testing is also built into the project, it could not be on anything like the scale of an entire desert, so it would be problematical to generalize from the results obtained there either. Even technically feasible geoengineering projects come with some very worrying unknowns attached, as the algae enhancement project also reveals. The experiments so far conducted on this have not yielded particularly promising results, and some scientists have dismissed the idea outright, pointing out that iron helps to produce methane, which is just about the last thing we want more of in the atmosphere, given its potency. There also seems to be some doubt as to whether the algae really would all sink to the ocean floor as expected; if they did not, there would be a danger of the carbon being released into the atmosphere again. It was thought that the more algae in bloom on the ocean surface, then the more carbon that would be transferred to deeper waters out of harm's reach, but marine scientists have now discovered that less transfer happens during a summertime bloom than in the rest of the year. Encouraging greater growth might actually have an adverse effect on the ocean's ecosystem: another worst-case scenario we could do without.

A way round the dangers posed to the ocean's ecosystem would be to cultivate algae on land, and one suggestion has been to turn desert over to this – raising yet again the issue of what this would do to the area's fragile ecosystem. One way or another, deserts as we know them look to be turning into an endangered species, mere sites for geoengineering exercises.

Renewables: The Downside

As discussed in Chapter 6, solar power has become a going concern in the American south-west, where the conditions for producing it are ideal – high sunshine levels, low population levels, and a scarcity of cultivable land (itself a testimony to environmental carelessness, however, as for several centuries it was host to the highly developed and relatively populous Anasazi civilization, until overcultivation and deforestation rendered it uninhabitable[7]). It has been claimed that as much as 250,000 square miles of the region could be turned over to the production of solar power, so this could be a really major operation that would go a long way towards meeting America's energy needs: a 'solar grand plan', as proponents have described it.[8]

Anything that affects the environmental balance of wilderness areas ought to be considered very carefully, however, and it is highly likely that a vastly increased human and machine presence will do just that. Wildernesses have their own unique ecosystems and when these are upset there are often unforeseen negative consequences further on down the line – sometimes well outside the wilderness area itself. The renewable energy systems are certainly very intrusive, and can be outstandingly ugly to look at, scarring what is otherwise a dramatic landscape quite badly. Some enthusiasts claim to find the effect pleasing, with one remarking on the SolveClimate website of a photograph of a central receiver plant: 'Notice: no smokestacks; no coal chutes; no rail lines stretching to the horizon for coal trains to approach. It's a beautiful sight.'[9] Despite that endorsement the plant in question is unmistakably industrial in appearance, although isolated as it is in a vast desert landscape the effect could be worse. To generate enough power to replace fossil fuels, however, we would have to assume a landscape almost completely filled up by such plants, which would be a different matter entirely – and one only likely to appeal to enthusiasts (and the owners of the plants too, of course, given the substantial profits on offer). The location of such systems in the American south-west has been justified on the grounds that they are 'on land that has few if any alternative economic uses', which is a depressingly narrow perspective to adopt, as if everything had to be judged primarily by economic criteria.[10]

There are plans in hand to build the world's largest solar power station on a 1,900-acre site in the Arizona desert in 2011 (although as I write it has yet to receive the go-ahead from the American government). This gives us some idea of the way the industry means to develop in the future, and it is inconceivable that such massive sites will not have a detrimental effect on the region's plant and wildlife. The intrusion of humankind always has this effect, but it has never occurred on this scale in such areas before. There is a tendency to plunge into schemes like this without thinking through such considerations (the profit motive playing a large part in that), or acknowledging that we do not really know what impact any alteration on the ecosystem in one area will have on others; there are subtle connections in the natural world that are not well understood by us yet – if understood at all. Nature is not a series of discrete systems, but an overall whole with a mass of complex interactions which can be affected by even small changes at any given point. What happens in the desert has an effect on what happens in the farmland and the cities.

As the comments above suggest, aesthetic considerations also arise when wilderness is turned over on a large scale to energy production. Much of the Earth is very crowded and the appeal of desert spaces is well established as a reaction to this, constituting areas where we can escape the stresses and strains of urban living, particularly as experienced in the world's major cities. Noise pollution is becoming a major problem globally, and desert areas are one of the few reliable refuges from it that we have left. Wilderness in general has had a powerful effect on the creative imagination, particularly in modern times, encouraging reflection on our relationship with nature, as well as on the power of nature itself; the sense of the sublime that nature in the raw gives rise to has inspired many a creative artist from the Romantic period onwards. It has to be acknowledged that something important would go out of our lives were we to lose a significant portion of what wilderness remains, and with it a sense of there being more to existence than the competitiveness of the urban rat-race.

A substantial amount of wilderness remains, but it is increasingly being viewed mainly as a prime site for the production of renewable energy: solar and wind in the main. The idea that the planet is there purely to service our energy needs is not an attractive one – whether or not you are a proponent of the Gaia concept. There have already been proposals put forward in all seriousness to grind down large areas of the Rocky Mountains to extract the oil contained in their shale, so we can see how far some energy companies are willing to go in their pursuit of product. It is bad enough that the far north of Canada is being badly scarred by the scramble to extract oil from the extensive tar sands there in Athabasca, but it would be an act of unforgivable environmental vandalism to subject the Rockies to the same kind of treatment on any systematic basis. Whether the peak oil crisis can be averted by other means remains to be seen, but the environment seems unlikely to come through it totally unscathed.

Solar power still has its problems in terms of its use on a worldwide basis. It is fine when it is located in desert areas such as the American south-west, with abundant sunshine to drive the system, but something else again in Northern Europe, with its grey skies and frequent rain (although, as we have seen, Germany has embraced it enthusiastically and with reasonable success). Weather is a critical factor in this form of energy production, and any scheme to reduce the amount of solar radiation reaching the Earth would not necessarily help either – especially if it were to go badly wrong in any of the ways I will be discussing later.

Wind power is not something new; windmills have been a major source of power for most civilizations until relatively recently. A wind farm is simply a more technologically advanced form of the windmill, although of course collectively such farms operate on a vaster scale. Both wind farms and projects to harness wave power can have a detrimental effect on the environment. Initially, the objections from the public were mainly on aesthetic grounds, complaints about spoiling countryside views, the same kinds of argument that were heard when electricity pylons were first introduced. (Enthusiasts have been known to suggest that wind farms could even turn out to be tourist attractions; yet why a dense mass of towers with huge propellers would be any more attractive than a line of electricity pylons will remain a mystery to many of us.[11]) But the protection of local ecosystems is increasingly cited as a reason not to proceed with the building of wind farms (I will be considering a particular high-profile example of this later in the chapter). There are also technical problems with regard to harnessing either wind or waves, and there are limitations to their effectiveness as providers of power on any mass scale – although the technology will no doubt improve as more research is done.

Like solar power, wind power is very much weather-dependent. Wind speeds have to average 25 kilometres an hour to be effective in driving the turbine, which places restrictions on where they can sited. Coastal areas, often out into the sea itself, are popular, with rounded hills and wide plains also favoured. Since the towers can be several hundred feet tall they can very easily dominate the landscape, which has been a cause of public protest. They can also be very noisy from the action of the propeller blades, although that should be less of a problem as the design is fine-tuned. If significantly more, and larger, farms are constructed, however, the collective mass might still be a source of noise pollution to anyone living nearby (as can happen particularly on sea coasts in countries like the UK). Wind farms can also be a danger to birds, who can be killed by the whirling propellers, especially when they pass by in flocks. Again, scale is important: the more wind farms, the greater the danger to wildlife.

Wave power is the least developed of the main renewables at the moment, and since it too is wind-dependent it suffers the usual problems of intermittency. Waves need wind to form and that is not always available. The waves also have to be fairly strong to drive the turbine, and that narrows the likely sites down. On the other hand, bad weather can disrupt the operation of the system, so striking a balance between weak and overly strong waves can be an operational problem. The Severn estuary in the UK, which has a significant tidal bore (6 feet high), has been identified as a possible contender for a wave power system, but not all estuaries would be as suitable. Remoter sites have already been exploited, however, with a successful power station – 'Limpet', standing for Land Installed Marine Power Energy Transmitter – currently working on the island of Islay, in the Scottish Hebrides.

Some systems involve equipment positioned on the sea bed, and that raises the possibility of altering the surrounding ecosystem, which as we have noted before can have unpredictable consequences.

If that method became widespread there is every likelihood that it could have a detrimental effect on the ecosystem around sea coasts, which could affect fishing and the farming of seafood. There is an aesthetic perspective here too; sea coasts of outstanding natural beauty could be disfigured by the construction of large-scale wave power stations, and that is the kind of situation that, yet again, can generate public outcry. Operational noise can also be a problem, although this can to some extent be masked by the noise of the waves and wind themselves (more so than in the case of wind farms).

Biomass is another renewable that been attracting increasing attention as a technologically very straightforward source of energy (through burning), and the idea of using the planet's waste materials would appear to have much to commend it. Rubbish and manure can be used, as can plant waste – sugar cane pulp, for example. As a culture we produce a massive, and steadily growing, amount of rubbish (the West being by far the worst culprit thanks to its cult of conspicuous consumption), and if it can be converted into fuel then that seems an eminently sensible solution to the multitude of rubbish mountains piling up around the globe. Collecting the basic materials for the plants to burn can be a more difficult exercise compared to other renewables, however, and the supply lines are far less predictable. A more serious problem is that biomass creates greenhouse gases, so it is not a clean fuel: renewable, yes, but that is its only selling point. (It should be noted that nuclear power is not a renewable source, although most commentators think we have enough supplies of uranium and so on, to last for some considerable time yet.)

Biofuel: The Downside

The effects of any massive shift from food to biofuel production – rising food prices; food shortages; social and political unrest, particularly in vulnerable Third World economies with much of the population living at or near subsistence level – need extremely careful consideration. The United Nations' World Food Programme (WFP) has even warned of biofuel production setting up the conditions for a 'perfect storm', as regards food prices, that poorer nations are in no position to withstand.[12] There have already been riots over rapidly increasing food prices in such marginal economies as Haiti. Even the West can feel the pinch over this; food prices in Britain, for example, rose at three times the rate of inflation in 2007, and if that trend continues it will cause problems in a developed no less than a developing economy. Concern is already beginning to be expressed over this, and it will no doubt turn into an important political issue quite soon. Although there is a complex of reasons for current rises in food prices, biofuel production is clearly a contributing factor – and will become even more so in the next few years

A switch to biofuel merely seems to encourage existing bad habits in terms of continued excessive car use and transport policy in general, thus deflecting attention away from the underlying problem of the systematic overuse of natural resources. As the global population rises, this problem can only become more acute, and unless it is addressed we are simply deluding ourselves that there is a quick fix that will magically make everything alright so that we can continue as usual.

The scale of the shift from food to biofuel production in places like the American Midwest is certainly impressive. One commentator has spoken of how farmers there are 'turning the corn belt of America from the bread basket of the world into an enormous fuel tank'.[13] The American government seems determined to encourage this trend, since the country's appetite for fuel is undiminished. The more attractive that biofuel begins to seem as a cash crop, then the more farmers will move into it. That is the way the market works: follow the money. Given the harsh treatment that many farmers have suffered in recent years at the hands of the supermarkets, who have driven their profit margins down to often barely economic levels in countries like Britain, it is not to be expected that they will resist the lure that biofuel is offering them at present. Using farmland to grow fuel could be seen as bordering on the unethical, but the market has never been particularly strong on ethics.

Similar efforts to boost biofuel production are being made in other countries, most notably India and Brazil, where the national governments are very actively promoting the biofuel option, both as a solution to their fuel needs and as a cash crop to sell at high prices on the world market. Sadly, in the former case particularly, this has led to massive deforestation. There is more than a little irony involved here when we think of the Socolow wedge that recommends that an area the size of India be planted as forest, when we are deforesting Brazil at such a rapid rate in the interim. Global coordination is sadly lacking.

Biofuel may in fact in many cases increase rather than decrease the quantity of carbon emissions, with deforestation a major factor. Everyone who is engaged in environmental modelling agrees that the world needs more, not fewer trees. Deforestation makes a mockery of the concept of offsetting, as if we had to increase our efforts massively in this direction just so as not to fall further behind – never mind to improve the situation for future benefit, the original motivation. Then there are the transport costs required to move biofuel from the often remote areas where it is grown to the main centres of population where it is most in demand. Add on the use of fertilizers (which can have the effect of releasing methane from the soil), and biofuel begins to sound emission-intensive in its production, thus defeating the whole point of the exercise. There are good and bad biofuels in this respect, and some create far fewer emissions in their production than others, but at the minute there is very little distinction being made and the market is accepting pretty well anything that it is offered from this source – without some regulation and guidance one would hardly expect it to do otherwise.

There are beginning to be some doubts expressed about the green credentials of biofuel from within the EU and some other national governments, so this is an area to be watched with some interest; regulations could well change as new evidence comes on stream. At the very least, biofuel will come under more scrutiny than it has in the recent past, when it was treated as one of the potential saviours for our transport energy needs. But the biofuel industry is becoming a substantial one and will no doubt feel motivated to fight its corner with some vigour. The market does not give up easily on proven sources of profit, and at the moment biofuel easily qualifies as that.

Biodiversity: The Health Factors

Climate change poses a severe trial to the world's biodiversity, with various species threatened by extinction as the temperature remorselessly climbs. Even if we accept Lomborg's evidence that polar bears are in a less parlous position than many of the warmers have been claiming, we would have to say that the long-term outlook in a rapidly warming Arctic is not very promising for that species – or indeed for any of the Arctic's denizens. (Reports of suspected cannibalism amongst polar bears cannot help the coolers' argument much either, one would imagine,[14] although this would require more authentication before we jump to any hasty conclusions.)

We can feel sad about the progressive reduction in biodiversity on both an aesthetic and an economic level, but another interesting twist to be added to the debate of late is that biodiversity, probably unknown to most of us, also plays a crucial role in human health. Polar bears, for example, are, according to a recent large-scale enquiry, the most useful source we have for the study of obesity-related diseases, thanks to their ability to gain or lose body fat quickly with no apparent adverse effects on their general health.[15] That is a fairly specific problem which might be resolved in various other ways yet (perhaps by an increasing global scarcity of food thanks to warming, ironically enough); but more worrying than that is the fact that when ecosystems are altered substantially this can create conditions that enable diseases to thrive as rarely before. As was noted earlier, many tropical diseases are already moving to higher latitudes owing to changes in the environment in recent times, so we have a precedent as to what we can expect on this score. It is not a good prospect for our collective health services – and yet one more reason, if it were needed, to take action to preserve the habitat of the increasingly highly symbolic polar bears.

Unfortunately, many of the supposedly green and clean solutions to climate change also constitute a serious threat to biodiversity. Wind farms, the harnessing of solar or sea power, the re-routing of rivers to overcome chronic water shortages: all such activities have an effect on the local ecosystems, often a quite devastating adverse effect. A plan to introduce a large-scale wind farm on the Isle of Lewis in the Scottish Hebrides, for example, has recently been turned down on the grounds that it would destroy the habitat of the local wildlife (particularly the bird population). Biodiversity won in this instance and no doubt Lewis will become a widely cited case as other such schemes are mooted elsewhere – as they certainly will be. But the argument back will be that we must of necessity develop sources of clean energy and that wind farms have to be located somewhere, preferably in remote areas where they can take full advantage of the elements and thus be at their most effective (wind not being the most reliable of sources, and needing to maximize its potential as much as it can). That argument will often win, and probably the more so as the effects of climate change become more evident and the demand for non-fossil fuel-based energy systems ever more insistent. One way or another, biodiversity looks set to be subjected to considerable strain for the foreseeable future, and we must now remember to include health on the list of things on which this development will impact.

Having worked through the problems, the projected solutions, and then the consequences that could arise from implementing the solutions, it is time to consider how we might reconstruct our geo-political relationships and narratives to cope with the situation we find ourselves in as regards the complex, and often paradoxical, interaction of global warming and globalization.

Part IV

Reassessing Global Priorities

10
Reconstructing Geopolitical Relationships: The Ethical Dimension

The relationship between the West and the Third World has to be rethought if we are to deal with global warming effectively. Whether one buys into the Gaia concept or not, it is clear that we are all in the same Earth system, and in real terms that means no one is immune from the impact of climate change; that surely has to be in the forefront of everyone's mind from now onwards. Jared Diamond makes a thought-provoking analogy between the plight of the Easter Islanders, whose remoteness meant that they had no escape route to fall back on once they had deforested and generally stripped their small island of its natural resources, and humankind as a whole, trapped on an isolated planet undergoing the same process. As for the Easter Islanders, it is a harrowing tale well worthy of the description 'apocalyptic'. With irreplaceable resources rapidly dwindling, as regards both food and materials, their society was plunged into civil war and even the practice of cannibalism in a desperate bid to survive. The population all but died out, leaving only a sad remnant of a once vibrant culture when European explorers arrived there in the eighteenth century. But unless we believe in the existence of friendly aliens we cannot expect an equivalent rescue package for the entire planet.

Diamond sounds a warning note based on the experience of Easter Island as well as various other Polynesian islands where existence was fairly marginal (many of them now completely uninhabited): 'When people are trapped together with no possibility of emigration, enemies can no longer resolve tensions by moving apart. Those tensions may have exploded in mass murder.'[1] The message is clear: either we cooperate with each other as much as we can, or we too could go under. Collapse can occur anywhere and at any time, and even the most complex society has its fault lines. (Ever one to see the brighter side of things, which I concede is not necessarily a bad trait to possess, Bjorn Lomborg reminds us that for all the 'irresistible image' it has for environmentalists, Easter was one of only twelve Pacific islands that underwent societal collapse and that there are around 10,000 others, many of which have hosted prosperous societies right up to the present.[2] The glass is always at least half full for this thinker.)

A change of consciousness will be necessary in the non-Western world as well as in the West, which will present a very considerable challenge to all the globe's cultures. Somehow or other, we must also keep alive the fact of cultural diversity and cultural difference through all of this, and accept that no one ideology has the answer to the crisis we are facing; the time for ideological point-scoring is well past. The West, after all, has led the way in environmental damage and the production of carbon emissions; but it achieves little to indulge in 'West-bashing' from a non-Western perspective, because we must all suffer the consequences of climate change, not just the rich nations. (It has to be remembered, too, that most nations with low carbon emissions are desperate to develop themselves economically, and the more successful they are at doing so, then the more their emissions total will rise.) The critical point about politics in an age of carbon footprint wars is that it must be truly global, and that is more than just a cliché in this instance. Social and economic policy must factor in not just their effect on global warming, but possible unintended consequences as well – and then strive to avoid those unintended consequences wherever they may occur. This means that we have to move past traditional notions of left and right and be much more flexible in our politics than hitherto; as opposed to 'flexible labour', 'flexible politics' is an entirely desirable concept to promote.

It will not be acceptable merely to curtail globalization; it will be necessary to find new ways to conduct world trade, or to provide substitutes for the benefits it previously brought into the Third World. Developing the Third World so that it has its own internal markets to sustain it, geared to its particular cultural history, rather than simply exploiting it as a cheap source of production for the West (outsourcing as the new colonialism), seems a direction in which it is worth heading. That really will require a radical change of mind-set as globalization is currently envisaged, particularly from the American side, which benefits more than anyone from the present set-up, but it has to become a priority. The priorities of global politics in general will have to be reassessed if we are all to survive; the West cannot act unilaterally without putting itself at risk of a mass migration from the Third World that would be to neither side's benefit, socially or economically. Again, we cannot afford to let this lie indefinitely; the sooner we start to address the problem seriously, the better.

The business world, too, has to be made to recognize that it cannot act in an independent manner, as if globalization had no geopolitical implications – or as if shareholders' interests alone were all that mattered when deciding on what course of action to take or what market to exploit. Corporate social responsibility has to become standard practice, and it has to amount to more than just a cynical 'greenwash' designed to dupe the public while 'business as usual' is prosecuted behind the scenes, or reinstituted at the first sign of falling profit margins and the panic this creates in the boardroom. Playing the shareholder card can no longer be taken as an unchallengeable justification for commercial practice; business ethics will have to change dramatically to take far more account of the (global) public good. It has to be realized that putting profits to shareholders before maintenance of the environment amounts to a form of violence against humanity, and that CSR is a condition of company existence rather than an optional extra that can be picked up or dropped on the mere whim of an executive board. Cynicism can no longer go on being the default position for business strategy, nor the current set of balance sheets the arbiter of business policy on the environment. Even the shareholders themselves will not benefit in the long run if the environment starts to collapse around them; no one gained on Easter Island.

Ethics: West to East

It is undeniable that the West has strong ethical obligations towards the developing world, where its colonialist past has left such a problematical legacy. Calls for international action to save the planet can be regarded with mistrust in the non-Western world, the suspicion being that these are generated mainly by self-interest and a desire to protect the Western way of life (which generated the problem we face in the first place). The West has to convince the rest of humanity otherwise, but there is no denying it has many hurdles to overcome in doing so after its treatment of non-Western societies in recent centuries. Edward Said's ground-breaking study of West–Middle East relations, *Orientalism*, set out to show how the West had constructed a notion of the East as a mysterious, essentially amoral territory, that had over the years, often under colonial rule, come to be accepted by the East's inhabitants as a true picture of life there: 'The Orient was almost a European invention,' as Said puts it, 'a place of romance, exotic beings, haunting memories and landscapes, remarkable experiences.'[3] The legacy this has left in the Middle East, the focus of Said's analysis, is still with us today in the region's deeply troubled politics, which show no sign of being on the verge of resolution despite years of intensive negotiation. Said's thesis can also be applied to the wider context of colonialism, with the West leaving a very similar legacy behind it in Africa, Asia, and South America.

That legacy has to be overcome if we are to address climate change efficiently as a global community, and the West must be prepared to look beyond national interests. Given that globalization is so much in the West's favour, it is imperative that it is reorganized to bring much greater benefits to the Third World, as hard evidence of the West's good faith. There has to be a check on what the multinationals can do, some code of conduct they are required to follow – plus some form of sanctions if they do not act in an ethical manner as regards working conditions and environmental impact, no matter where on the globe they are operating. Too often, as Fred Pearce has only too correctly observed, products are sourced from the developing world in an 'ethical vacuum', and that just has to change.[4] As Pearce himself demonstrates in *Confessions of an Eco-Sinner*, that means all of us should be asking a lot more questions about the means of production behind 'where our stuff comes from'; otherwise we are all complicit. The anti-capitalist movement is already asking such questions, and if it is a rather blunt instrument of change, it is still a necessary part of the mix.

Ethics: East to West

The East also has ethical obligations back to the West. Islamic countries, for example, need to work to create conditions where militant Islamism does not flourish as it is doing in several cases at present. Like any religion, Islam can be, and over the course of its history has been, interpreted in either a moderate or an extreme way, and the moderate aspect should be encouraged as much as possible. The West alone cannot do this, although it would help if it ensured that conditions were made as receptive as possible for the expression of moderate Islam within its own territory; but the main impetus will have to come from within the international Islamic community. Unless such a change of attitude occurs, the mutual suspicion that currently exists between the West and the Islamic world will continue to sour global political relations, and with that the prospect of real progress towards global cooperation on climate change.

As peak oil approaches it becomes even more important to reconcile the West and Islam. Much of the world's current oil supply comes from Islamic countries, and we ought to be thinking beyond that point, when a fall in revenues in those regions potentially will exacerbate social tensions there that could be exploited by zealots for religious purposes. It could be a test case for what a retreat from globalization would be like. The Islamic world stands to be hit very hard by a move away from fossil fuel – whether that is forced on it by circumstances, as in the end of the oil reserves themselves, or because of a shift into cleaner fuels in the West. How the main countries involved handle that transition could be critical for global political relations, and the West will have to be as supportive as it can, while recognizing that it cannot dictate what will happen within those countries. Everyone must be committed to improving the state of international relations; otherwise we have little hope of being able to address the various crises that global warming is preparing for all of us, whatever our ideological persuasion may be.

An example of a change of attitude within the Islamic community that the West can take heart from is the recent declaration against terrorist activity by Sayyid Imam al-Sharif (better known in radical Islamist circles as Dr Fadl), published in the Egyptian and Kuwaiti press in 2007. Fadl's earlier book, *The Compendium of the Pursuit of Divine Knowledge*, had on the other hand encouraged Muslims to take violent action against non-believers, and had become something of a handbook for indoctrinating recruits into movements such as al-Qaeda. There is still a radical streak to his work, and Fadl has not entirely rejected the notion of jihad to assert Islamic principles, but he has certainly moved away from the policy of indiscriminate violence that has inspired radical Islam of late. Material such as this needs to be circulated as widely as possible and used as a basis for debate.[5]

Corporate Social Responsibility: Overturning the Greenwash Mentality

How can CSR be turned into something more than just a greenwash designed primarily for public relations purposes? There has to be more officially led pressure on the business world to demonstrate that they will put the public interest first and are willing to accept limits on their pursuit of profit. Environmental costs have to start being built into company planning, and where these are seen to be significant, or problematical in any way, then the activity should not be pursued – or perhaps should even be banned if its consequences seem dangerous enough. Such a policy would counsel companies away from clearing peat swamps, or engaging in large-scale deforestation programmes that simply release more and more carbon into the atmosphere – to everyone's ultimate detriment, as is now abundantly clear. Or, for that matter, searching for loopholes to enable them to circumvent legislation which has been put in place specifically to reduce carbon emissions: shipping biofuel back and forward across the Atlantic as a notorious case in point, just in order to shave a few points off the price. The profits that are made from such activities are at the expense of humanity, and that should be broadcast. The environmental lobby (Greenpeace, EarthFirst! et al.) has been doing just that for some years now, but they only impinge on the public consciousness up to a point – and their politics are not always to everyone's taste. But government-sanctioned pressure (through its own monitoring agencies, for example) would register much more powerfully, since the government and its agencies are both publicly responsible and publicly accountable; at least in principle, all of us have a stake, and a voice, in the policies they implement.

Companies should be required to give regular accounts of the environmental costs their activities give rise to, and these should be widely circulated in the public realm to identify abuses before they get out of hand. Circumstances surely call for such preventative measures to be taken wherever they possibly can be. There ought to be much more auditing of business to ensure that it is not putting profit before the environment, and the state has to act for the general public in this respect, insisting on compliance with agreed standards – and preferably globally agreed standards. Neither will it be enough to claim that offsetting can cancel out those environmental costs; those arguments no longer carry much weight, not now that we know trees are no longer able to perform miracles on our behalf. That has to be seen as part of the greenwash mentality, and to be challenged vigorously. There is no easy way out of environmental vandalism – which is the only appropriate way to describe mass deforestation schemes, wherever they take place, or any other project which releases huge amounts of carbon or methane into an already dangerously over-polluted atmosphere. The ethical vacuum that so many companies operate in has to be made a thing of the past.

The ethical dimension to reconstructing our geopolitical narratives, including the relationship between business and the community, seems very clear, and it ought to be leading us to explore new methods of collaboration and cooperation as a matter of some urgency. The theories that could be drawn on to make these methods as effective as possible constitute the next topic.

11
Reconstructing Geopolitical Narratives: A Radical Democratic Globe?

Our old political narratives are in desperate need of redrafting, and the left in particular will have to work out a much more flexible approach to global politics, such that unintended consequences of the kind that have been discussed over the course of this book are minimized. The Green movement, which sometimes can be very reactionary indeed in its thinking, should also be reconsidering its narratives and their relevance to the global economic system. It is time for some non-linear thinking to be applied to our non-linear world. The extent to which the ideas of 'radical democracy' – as developed in the first instance by the political philosophers Ernesto Laclau and Chantal Mouffe – could help develop such a new form of political consciousness, and thus promote a fresh outlook on global priorities in an era of accelerating climate change, is worth exploring. Radical democracy was devised as a response to the decline of the traditional left over the latter part of the twentieth century (with the collapse of the Soviet Union just around the corner accelerating the process). It recommended that there should be a plurality of voices contesting the political terrain, none of which should be silenced as long as they accepted the rules of engagement – open-ended debate, no attempt at quashing opposing views, respect for one's opponents no matter how much one disagreed with their position, for example. While this can be notoriously hard to implement in the two-party political system that tends to apply in most Western democracies nowadays, it makes more sense in the geopolitical context of competing nation states.

The problem at the moment is that the globalization approach is based on a 'one size fits all' model, with the West deciding the size, whereas it would be far more sensible for there to be a variety of economic models reflecting each country's particular cultural and historical circumstances – and for all involved to respect these differences, even to seek to preserve them for the global public good. We have coped with this before for most of our history, so why can't we do so again now? Is it really so problematical if some countries engage in a certain amount of protectionism? As long as this is kept within bounds, surely the rest of us can learn to live with it? (Paradoxically, or hypocritically enough, America, that great champion of the free market, is still capable of indulging in protectionism when it suits it to do so – as in the case of its steel industry.[1]) At the very least this calls for a fundamental change in attitude in institutions such as the World Bank and the IMF, which still tend to favour ideological purity over cultural difference, as if homogeneity in economic dealings was the highest virtue to which humanity could aspire.

Granted, it is easier to transact business if everyone is following the market fundamentalist line, but why should we be so obsessed about making life easy for business? If there are barriers to surmount, different systems to come to terms with, or new cultural reference points to be learned, then business will find a way if it has to; it will require more effort and constant changes of approach and direction, but that seems a small price to pay to maintain cultural difference. Globalization does not have to mean homogeneity, the dreaded McDonaldization of culture that so many complain about, although that tends to be the way it is currently interpreted. As the philosopher Julian Baggini has pointed out, '[t]he free movement of goods, like the free movement of ideas, might just work better when there is more choice available.'[2]

Radical Democracy

Laclau and Mouffe's *Hegemony and Socialist Strategy* called for a shift to what the authors dubbed 'radical democratic politics', where a much wider range of opinion was to be included than is usual in the current political set-up in the West.[3] These ideas have since been developed further to show how they might work within the Western democracies in order to correct what Mouffe has described as the 'democratic deficit' that tends to occur in most cases, with minority voices being marginalized by the mass political parties, which have the power and the funding to dictate the political agenda of their nation.[4] We might claim a similar deficit in terms of international relations, with the concerns of the developing nations all too regularly being marginalised – plus a very substantial economic deficit when it comes to globalization, which in the current free market system is plainly constructed for the West's benefit. The reaction to global warming is only likely to widen such deficits, to the obvious detriment of the developing nations collectively; if we want true international cooperation these have to be reduced.

Mouffe puts forward the concept of 'agonistic pluralism' as an ideal in the political realm, distinguishing this from antagonism, where neither side really accepts the right of the other to exist and is concerned instead to overcome its opponent and destroy its public standing – Lyotard's differend in action, with neither side really listening to the other, or according it any great degree of credibility. Mouffe identifies a 'dimension of antagonism that is inherent in human relations', and takes it to be the goal of radical democracy to overcome this in order to open up political debate to a wider constituency.[5] A further problem in Mouffe's view is that the two-party system is very often a consensus which excludes anyone outside quite a narrow band of opinion that both parties are prepared to tolerate. This arrangement may give the appearance of choice and oppositional debate, but in most cases it is mere illusion. The differences between the two main – or even three, where this happens – parties are more a matter of style and presentation than of real substance; there is a dominant ideology in operation restricting the scope of debate. Agonistic pluralism, on the other hand, assumes a constant round of searching debate between a wide range of groups, each trying to convert its opponents to its own position rather than working to reach a compromise with others. Compromise is in fact strenuously to be avoided in the radical democratic scheme, being viewed as a sell-out of one's beliefs. Certain conditions are laid down, such as a shared commitment to democratic principles and a recognition of the right of others to enter into the debate, but otherwise the more spirited and contested the discussions are, the better. Antagonism is what results when this kind of forum is not available and many feel left out of the political process, their viewpoints unrepresented and concerns marginalized (as can happen so easily with the environmental lobby, whose message can be very uncomfortable for the dominant ideology to absorb). The result is often violent conflict, born out of frustration.

The point of agonistic pluralism is to ensure that all voices have a proper chance to be heard in the political arena, thus serving to narrow the democratic deficit. How viable a method of conducting politics this would be in the average nation state as presently constituted in the West is another matter, and it has to be admitted that it would require quite a dramatic change of public and political consciousness to bring this about, so entrenched is the current system. As David Howarth, a political theorist very sympathetic to the notion of radical democracy, has observed, 'less attention is paid to the economic, material and institutional obstacles that block its realisation, as well as the precise composition and configuration of such impediments.'[6] There is, as Howarth goes on to argue, 'an institutional deficit' in such thought which still needs to be addressed.[7] Nevertheless, the idea of widening political participation within a non-threatening pluralist framework, where difference and diversity are both respected and fostered, seems worth pursuing. What is required is a shift to what Howarth and his writing partner Jason Glynos call a '*problem-driven*' system of debate, 'rather than *method-* or purely *theory-driven*'.[8] Such a system is by its nature much more fluid and, so the argument would have it, more responsive to political problems – and climate change certainly needs that heightened level of responsiveness from the global community.

The more we blur the boundary between right and left in political terms, however, the greater the danger of lapsing into what Mouffe derides as 'post-political' thinking, in which a new consensus is reached, to the detriment of the agonistic approach.[9] Mouffe is critical of this move away from partisan-based politics to a situation in which, '[i]n place of a struggle between "right" and "left" we are faced with a struggle between "right" and "wrong".'[10] While I think that Mouffe is correct to insist on a partisan dimension to politics, I also feel that climate change (which is not really one of her concerns) very often is more a question of right or wrong than right or left, and that we must face up to the implications of that if any progress is to be made on this front.

Think Glocal

Ulrich Beck's ideas on cosmopolitanism are worth exploring in this context, since they too call for a reconfiguration of democratic political debate. He urges us to move to what he calls the 'glocal' dimension, whereby we move past the confines of national politics as they are currently practised (although, as we shall go on to discuss later, for Mouffe glocal equals post-political). National and global politics are now inextricably mixed and that demands a different approach than hitherto. We need a 'transnational framework' in which to pose glocal questions properly, and for this to come about 'there has to be a reinvention of politics, a founding and grounding of the new political subject: that is, of *cosmopolitan parties*.'[11] Beck sees himself as putting forward a 'Cosmopolitan Manifesto' for the new situation we find ourselves in, where 'the central human worries are "world" problems . . . because in their origins and consequences they have outgrown the national schema of politics.'[12] That very aptly describes the condition of a world facing global warming, where carbon emissions from anywhere become a problem everywhere. There truly are no boundaries involved when it comes to climate change; it is constitutionally transnational.

In a world responsive to cosmopolitanism, Beck contends, '*risk-sharing* . . . can . . . become a powerful basis of community'[13] – which is precisely the point of the *Stern Review* and its call for a globally coordinated response to the effects of climate change. 'Risk', for Beck, 'is the modern approach to foresee and control the future consequences of human action, the various unintended consequences of radicalized modernization.'[14] Community becomes the key, with cosmopolitan parties seeing themselves as representing the planet rather than just the narrow, provincial concerns of individual nations, which are all too often unwilling to take account of the bigger picture – a critical problem when it comes to climate change, as we have repeatedly seen. If America's bigger picture extends no further than America itself, then we are all in trouble.

Beck discriminates between globalization as we know it at present and cosmopolitanism, although he concedes that the former has created a sense of a world society amongst much of the population that cosmopolitans can draw on for their own purposes (just as economic globalization, in Naomi Klein's view, has brought about the possibility of a new form of socio-political globalization to combat the former's excesses). There is a global capitalist community, so there can also be a global cosmopolitan community. The basis for assembling the latter is there already, and the time is ripe for experimentation in this direction: 'why can or should the political be at home or take place only in the political system? Who says that politics is possible only in the forms and terms of governmental, parliamentary and party politics?'[15] Ultimately, Beck feels we need to develop a 'responsible globalization'.[16] Once again, as with radical democracy, we are being asked to break through current ideological constraints and adopt a more creative attitude towards the resolution of problems that transcend national boundaries. Kyoto was an opportunity to go glocal in the way Beck recommends, but it is instructive to note how nationalism, American in the main, undermined this. Yet it is clear that if we go on in that way, with a national interest-led rather than problem-led approach, global warming will eventually spiral out of our control.

Glocal should also mean a significant investment in 'green education' in all countries. Just as there should be encouragement to think across national boundaries and share risk, so there should be sustained efforts to educate the population as to why this is necessary and what it is trying to achieve on behalf of the planet. Green education should be built into the curriculum at school level from an early age onwards, such that it is made clear that environmental issues transcend national boundaries and require everyone to be receptive to the risk-sharing ethic, to possess a cosmopolitan mentality. Generations to come just have to be more knowledgeable about the environmental problems we face, and how to minimize them, than the majority of us currently are. Getting across 'the green message' in schools is, as one commentator has noted, 'likely to be the most effective, and efficient, way to green the population'.[17]

Mouffe is critical of Beck and the entire project of cosmopolitanism, arguing that it amounts to a new post-political vision which eradicates the partisan aspect that for her is essential to properly functioning politics. She wants adversarial politics; Beck, she claims, wants a broad consensus based on liberal democratic ideals. But without wishing to go too deeply into what is a complicated area of theoretical debate, I think we can take something from both sides in terms of the issue at hand: that is, climate change and the various ways we might deal with it – as well as the various ways we might conduct the debate over the options on offer. We can be partisan about globalization (concern about the social injustice it involves is not an attitude confined to the left, but it is certainly more prevalent there and more passionately expressed), while also recognizing that it would not necessarily be to the advantage of the developing world for there to be any significant withdrawal from the system. Perhaps it has to be conceded that if any action is to be taken on a phenomenon that, like climate change, is unquestionably international, then some kind of consensus has to be reached?

There is probably more common ground between the radical democratic and the cosmopolitan positions than Mouffe acknowledges, since even she insists on agreement – one could say consensus – on the rules of engagement: 'A democratic society cannot treat those who put its basic institutions into question as legitimate adversaries. The agonistic approach does not pretend to encompass all differences and to overcome all forms of exclusion.'[18] And Beck's cosmopolitan parties could embrace various shades of opinion in an adversarial manner easily enough. This could even happen in the green education enterprise, since there is no universally agreed policy on how best to deal with global warming. A range of viewpoints can be advocated; as long as there is debate going on, then consciousness is being raised in the manner that is required.

There is an argument for vesting more power in the UN, which, as theorists such as Danielle Archibugi have recognized, is as close as we come at present to a cosmopolitan organization: 'It is neither realistic nor useful to imagine a more democratic global governance without assigning a principal role to the UN. There is no alternative to the UN as such, and its reform is needed to allow for better use of the organization.'[19] The UN needs to function more like the EU, with those countries who sign up to it incorporating its rules and regulations into their national systems as a condition of membership, accepting that the transnational framework is ultimately in their best interests.

There are always problems when very large power blocs are created in the political sphere, the main fear being that they will become authoritarian and even totalitarian in their operation, leaving many of those they have control over feeling very vulnerable. There do need to be checks and balances in place in such cases, and these would have to be carefully monitored to prevent abuses of power occurring, especially if we are talking about a global power bloc. But it is not beyond the bounds of possibility to make such a system work for the common good. The EU has been astoundingly successful in improving the economic performance of all its member countries, and in keeping Western Europe war-free since its establishment in the aftermath of the Second World War – no mean feat in a continent where national rivalries have led to a host of major conflicts over the centuries. At its best, it fosters a European cosmopolitanism that has been instrumental in changing the face of European politics for the better. (On the negative side, one would have to say that the EU's trade deals with the developing world can sometimes be very exploitative, in line with the current globalization model where the developing countries have little real bargaining power with Western power blocs. That is an aspect of its operation which clearly calls for serious review.)

There are sceptics, particularly in the UK, where the popular press is notorious for its anti-Europeanism and does its best to stir up opposition to what it sees as a bloated bureaucracy blighting our lives. Yet the EU does not seem to have destroyed national identities (France is fiercely protective of its own, for example), despite its members coming to recognize the virtues of acting in the common interest and implementing laws decided upon in Brussels or Strasbourg rather than in their own governmental institutions alone. Its popularity can be gauged from its expansion to include the ex-Soviet East European bloc, who have been queuing up to join in the past few years. British sceptics still bemoan a perceived loss of national sovereignty, but the EU is surely a good model of a transnational organization, with far more to be listed on its credit than debit side. The European political narrative has been rewritten since the 1950s, and to everyone's advantage – one has only to consider the first half of the twentieth century to see what a massive achievement this has been. We can learn from such transnational success stories, where the national is able to retain a sense of its own identity within a loose, but effective and respected, federalist grouping.

We could use something like the EU system on a global basis to deal with climate change and globalization in an efficient manner: a body charged with responsibility for global energy policy and all its ramifications (including expediting and overseeing green education perhaps). This would not be world government on the now largely discredited centralist model (the Soviet, for example, with its demand for ideological unity across all cultures, regardless of their individual history), but a diffused government globally with agreed objectives and methods to deal with a specific problem: risk-sharing on a properly international scale. The current system, with no one in overall control, but some with far more power to impose themselves than others, serves no one's long-term interests. Our own lives are determinedly short-term, but human beings have always shown the ability to think further ahead than that and act accordingly – no one would much bother planting trees otherwise or, to go back in time, starting work on a medieval cathedral designed to take several generations to build. We have to appeal to that generous side of our natures if progress is to be made on a programme of risk-sharing for the sake of the entire planet. The more cynical of economic theorists notwithstanding, *Homo economicus* does not define us totally, neither as individuals nor as a species; we are capable of recognizing that there is more to life, and politics, than 'it's the economy, stupid.'

The State of Nature Versus Agonistic Pluralism

Most of the time the relations between nation states have more the appearance of Thomas Hobbes's 'state of nature' condition than of agonistic pluralism: that being a condition of constant conflict with no enforceable guarantees of personal security.[20] Every nation is motivated in the first instance by its own self-interest, and negotiates with others trying to protect and further this as much as possible, while being well aware that its own security is always tenuous and can be compromised at a moment's notice (as we have seen in the case of the USA and the 9/11 event, bearing out Hobbes's contention that even the strongest are at risk in the state of nature, where no one can put themselves completely beyond the threat of attack on a permanent basis[21]). National self-interest represents a significant barrier to dealing with global warming, acting as a constraint on the development of the global consciousness necessary to deal with the problem effectively; Mouffe's 'potential antagonism' is very much in evidence when it comes to relations between nation states, which can be very suspicious of each other, and that just has to change. A risk-sharing mentality has to be developed, with the appropriate global mechanisms to back it up.

Globalization in its present form merely encourages national self-interest, pitting nations, and particularly developing nations, against each other in order to capture the business of the multinationals, who are subject to little in the way of external controls as they move unhindered around the globe, locating themselves where they are able to negotiate the most advantageous business deal. Considerations of global warming do not loom very large in these cases – if at all. Multinationals can move on to another location whenever it suits them, and that threat is hardly an invitation to discuss the finer points of business ethics. Deals are generally done without much reference to environmental consequences; the agenda is unrepentantly economic. Sadly, there are always takers for whatever meagre offers the multinationals see fit to make, so desperate is the condition of so many developing countries; 'tragic choices' continue to be made. In so many ways it can seem like the colonial system revived, the power manifestly residing with the Western interest - regardless of what globalization defenders such as Martin Wolf may say. Even those non-Western countries which are currently experiencing the advantages of globalization, such as the oil producers, cannot hope to do so indefinitely. They too will discover themselves on the wrong side of globalization when peak oil arrives, and no matter what they may think, their position is quite fragile in the longer term. A move into solar power production is a possibility, but that will probably require even more in the way of international cooperation than oil does, not least in addressing the environmental consequences.

Combining the best of radical democracy and cosmopolitanism could help us overcome this state of affairs, and foster a new sense of common purpose on the subject of both globalization and climate change. This is not so much a new ideology as a change of perspective, in which the transnational takes precedence over the national: a new narrative to live by, with objectives that will benefit everyone. What we have to avoid at all costs is any lapse back into the old political narratives, and their commitments based on a very different world from the one we now inhabit – never mind the very different one again that is looming ahead of us in the form of frightening environmental tipping points. Debates about ideological purity will be of little consequence if we pass over those. Nations can no longer regard themselves as independent in the traditional sense of that term, not with the necessity of risk-sharing being so evident – that way, only unproductive differends lie. There still needs to be a range of competing viewpoints on offer, however; it is more a case of whether these are in agonistic or antagonistic relation to each other. In a world at risk there is no place for the latter.

Drawing on the spirit behind radical democracy and cosmopolitanism, as well as investing heavily in green education so that the next few generations will have a much more sophisticated grasp of environmental issues than the general public does now, our geopolitical narratives can be reconstructed, and that would give us a much firmer basis on which to tackle climate change and the various trade-offs this will require.

12
Conclusion: Survival, Disaster, Trade-Off

The point of this book has been to make it clear that as a matter of urgency we must start addressing the problem of how to engineer the best possible trade-off between economic survival and ecological disaster – and on a global rather than a national basis. Ultimately, it is to this that the carbon footprint wars come down. National self-interest, and the politics lying behind it, cannot be allowed to dictate how we go about this task; those days have to be put firmly behind us as we strive to keep the Earth system as stable as we can. Neither can we continue to assume that the rights of shareholders are always to be considered as sacrosanct – those days have to be put behind us as well, and the concept of stakeholder pushed much harder than it ever has been in the past. Financial power alone cannot be allowed to determine the value of human life or the destiny of the human race. I will now summarize the various suggestions made over the course of the book for reconstructing and revitalizing our narratives of global warming and globalization to that desirable end. The goal must be to keep the worst-case scenarios just that, scenarios only, which serve to concentrate the mind on how to hold our position at the edge of chaos rather than tipping over into collapse. As long as apocalyptics is a literary genre that exercises a fascination over us, a source of enjoyable frissons for the reader, that is all well and good; what we do not want is it turning into a nasty reality.

Carbon Truth

If we are to survive in something recognizably like our current form, then it will also be absolutely necessary for governments to be truthful about the carbon emission situation – and for there to be means of checking that they are doing so, ideally by some supranational body such as the EU or the UN. The IPCC should be given more power to insist on targets being met and procedures to monitor those targets; following their recommendations should not be left to national whim, nor to the devious tricks that governments are only too capable of playing with data. There is a very real danger that governments, as is their wont with future elections in mind, will tell the public whatever it is that casts them in a good light, and from now on that is going to mean recording reductions in national carbon emission totals – as a result of government policy and efficiency, we shall of course be told. For those who have signed up for Kyoto this would seem to be a test of their political credibility, but governments can be very creative with statistics, of which they have a vast amount to hand at all times. Bjorn Lomborg may put his faith in statistics as the way to provide us with an accurate picture of reality, but just recall how they are being used in the current Argentinian inflation debate.

Statistics can be manipulated to political advantage, therefore, and a recent event in the UK provides yet another instructive example of why we have to be on our guard. In spring 2008 the British press reported that the country's National Audit Office was challenging the government's figures that the UK had achieved a 16.4 per cent cut in carbon emissions since 1990. According to the government the country had released 656 tonnes of CO_2 into the atmosphere in 2005, but it transpired that there were in fact two official sets of records being kept on carbon emissions, and that the other one, from the Office of National Statistics, showed a significantly higher total of 733 tonnes released that year. Even worse, the National Audit Office considered the latter's records 'more comprehensive as they include aviation and shipping emissions'.[1] (Omitting shipping emissions is something of a global-wide tendency, and now that shipping has been shown to be a higher source of carbon emissions overall than air travel, a scandalous one as well.) Predictably enough, perhaps even cynically enough, it was the first set of figures that the UK chose to log with the UN as evidence of its seriousness about global warming through adherence to the Kyoto Protocols. More a case of risk-dumping than risk-sharing, many of us might be inclined to say.

One would have thought that if there was ever any issue not to play politics with then surely it would have to be this; the consequences of deceit are too high, far higher than the outcome of any mere national election or the reputation of any particular government or politician. True, in this case the government were caught out and made to look bad on the international stage, but how many other governments are playing a similar game, one wonders? And can we be sure that governments who are caught out, the UK included, will then mend their ways, as opposed to trying to find other more devious methods of hiding the real state of affairs from us such that they can claim the moral high ground for short-term political advantage? Just framing the question makes one suspect what the likeliest answer would be.

To show the games governments are capable of playing with this issue, the British government has recently also been lobbying the EU for changes over the way it sets its targets for the production of clean energy based on renewable sources. The UK has requested that it be allowed to include British investment in clean energy facilities elsewhere in the world – specifically outside the EU – in its own targets, leading the director of Greenpeace to remark sarcastically that, '[t]his would allow a UK minister to lay the foundation stone of a power station in China and say it counts as our contribution to European renewable energy targets.'[2] Another outraged spokesperson from the environmental pressure group World Wide Fund for Nature (WWF) saw this as yet one more depressing example of Britain 'trying to evade its environmental responsibilities' by striving to find loopholes in the interpretation of legislation.[3] The critical point, however, is that for any country to act in this manner is not just to evade its own responsibilities since it is the whole planet that will suffer eventually; it is an international rather than a merely national or regional concern. The temptation to regard the reduction of carbon emissions as an abstract issue – like inflation targets, for example – rather than a matter of survival is only too evident at such junctures.

Other governments will no doubt be observing this particular debate with some interest to see how successful the UK proves to be in bending the rules to its own temporary benefit. It is not a good precedent for any national government to set, and an indication that 'breathtaking cynicism' is not confined to the business community alone: a greenwash in one, a carbonwash in the other. One can only say that this amounts to ecocide by stealth.

I suggested in Chapter 11 that it would be a good idea to establish a body to oversee global energy policy. One of the duties of such a body should be to obtain and publish accurate figures about carbon emissions (based on unambiguous rules of measurement which it would formulate), and to promote best practice. The worst offenders are the more developed countries, but even in the West there are variations in emissions per head. The USA's figures are nearly double those of Germany, which as we have seen, has a significant investment in green energy, and the former could learn from the latter the benefits of a more integrated approach (France, with its heavy dependence on nuclear power has an even lower figure than Germany). But any body that was established would need to be given real power to impose targets, and sanctions if these were not met within a reasonable timeframe. There has to some ceding of national sovereignty on this issue; otherwise we run the risk of repeating the Kyoto stalemate. What is required is a Kyoto-style agreement that is binding, and that all parties have accepted beforehand, in the common interest, is binding. Unless the developed nations take some lead on this, then there is no incentive for the developing countries to be concerned about their own rising emission totals. While countries can learn a certain amount from each other as to how to reduce their totals, we can no longer rely on such a piecemeal approach.

Carbon and the Free Market

We need, as a matter of urgency, to reconsider our commitment to the free market if we are to reduce the size of our carbon footprint.

Nigel Lawson may believe that we should rely on the market for evidence as to what methods, if any, to adopt in emission reduction, but it is not only 'red is green' environmentalists who would want to contest that solution. Shareholders cannot be allowed to risk the existence of the planet in the name of increased share dividends, and that is always going to be the primary concern of the market. Playing the shareholder card can no longer be seen as an acceptable tactic for any company to adopt, because it is no longer the real bottom line. The free market has had many positive effects on social development, and its benefits to humanity are not to be underestimated (which the more radical environmentalist campaigners often do, whether they are of the 'red is green' persuasion or not, thus helping to reduce the appeal of the environmental lobby in general). Whether it is as integral to freedom as thinkers like Milton Friedman contend, is much more debatable, but there is no denying the market's success as regards raising living standards.

The market has had many negative effects on social development too, of course, and it is not the only way to organize our economic life – as any socialist will be quick to confirm. The other systems have in the main proved to be less successful in delivering material wealth on any mass basis, but that should lead us to consider whether quality of life is best measured by that criterion. It should make us wonder as well whether the material wealth is all that meaningful after a certain level of individual comfort is reached (Are we happier with two cars than one? Three?). Nevertheless, free market capitalism has come to be the dominant economic system of our time, and the collapse of the Soviet Union has, for the time being anyway, removed the last effective check on its operations from an opposing camp. This is even more the case since China has embraced free market economics (if not its usual political counterpart of parliamentary democracy, which might make us question Friedman's assumption of there being a necessary connection). The trend in recent decades has been to create a progressively less regulated market, and globalization has carried it to every corner of the world in that guise.

But the market cannot continue in that ultra-free mode without dragging humanity down with it. The complete freedom from constraint demanded by the market fundamentalist lobby cannot be reconciled with tackling global warming really seriously – or social justice either, come to that, as we can see from the fate of so much of the developing world. The market mentality is constitutionally short-term; left to its own devices it will invariably opt for quick profit to the exclusion of all other considerations (think how the stock market careers madly up and down at the slightest provocation, often little more than mere rumour), and profit with as little outlay as it can manage. 'The market has shrugged off any responsibility for democracy and society in the exclusive pursuit of short-term profit maximization,' in Ulrich Beck's trenchant assessment, leading to what this commentator has dubbed a 'free-market farce' that merely underlines our need for a strong state to watch out for us[4] (those who think the less government, the better, might be advised to look at countries where this applies – such as war-torn Somalia).

In effect, the market is putting us all at risk in the name of the shareholding community (an amorphous group, making it hard to campaign against, very much a faceless enemy). The kind of financial outlay required to put schemes in place to alleviate global warming is not going to be forthcoming from the commercial sector, which would balk at the lack of immediate returns – or any tangible returns at all, apart from the ability to continue trading more or less as normal, which does not improve balance sheets in the present as the system insistently demands. Growth is always the mantra, but it is a mantra that has to be challenged in the circumstances we now find ourselves in, with carbon levels at a 650,000-year high and rising rapidly. If economic growth means carbon emission growth, then we should be re-examining our assumptions about how our society operates and what we expect from it.

Western governments in particular will therefore have to be far more interventionist-minded than they have been in recent decades, when in most cases they have been all too willing to allow the market free rein, as its apologists have claimed is the best policy. Friedmanite doctrine has laid down deep roots in this respect. In retrospect we can see that this laissez-faire policy has served to drive up carbon emissions to their current record levels, as business has sought to maximize demand for its products however it could, and as quickly as it could. Any gesture towards the environment has tended to be of the 'greenwash' rather than the sincere variety: a means of diverting public criticism in the main, rather than an ethical commitment (although the public is beginning to see through such manœuvres). We may not be able to blame the early free marketeers for bringing us to our current pass, since they could not have known where their activities ultimately would lead in a global environmental sense; but we can blame those who refuse to heed the evidence now, and proceed onwards with a 'business as usual' approach, blinkered to the effects of their policies.

Fred Pearce thinks the tide is turning: 'I do detect big changes afoot. Many large corporations, especially in Europe, are now actively asking for governments to set tougher targets on greenhouse gas emissions.'[5] However, he does concede that it is mainly the profit motive that is driving this attitude, the hope of getting in on the ground floor of some new process or product and thereby gaining a critical advantage over one's rivals. Hence the approving response of Richard Clark, Pacific Gas and Electric's Chief Executive Officer (CEO), to the carbon trading system generated by the USA's Clean Air Act of 1990: 'The environment isn't just a money loser – it's a profit center.'[6] Gabrielle Walker and Sir David King make similar noises: 'Around the world, brokers and businesses have noticed the amount of money changing hands over climate change and are eager to join in.'[7] Yet we know that the profit motive is at best a problematical method of achieving the public good, and much more evidence would have to be forthcoming before absolving the corporate sector entirely of their many sins against the environment (which most of us find hard to think of as just 'a profit center').

As things stand, free market capitalism and planetary survival appear to be on a collision course with each other. Jared Diamond insists that we cannot expect businesses to act like charities, as that is not why they were developed; but even so we cannot allow CSR to be merely an optional add-on to organizational operations. It has to be standard practice, rigorously adhered to, and subject to strict regulation and constant monitoring; otherwise the baser human instincts such as greed will, sadly enough, almost always come to the fore. Greed is a fact of life, but its worst excesses can, and should, be tempered wherever that is possible. Even if we all conclude that we have to use the market model (for the foreseeable future, anyway), it can be interpreted in a variety of ways. There is nothing unethical about taking local needs into account and resisting the market fundamentalist drive towards uniformity.

If the left have had to come to accept the fact of the market – even if very reluctantly for some – then the right also must come to accept the necessity of there being limitations to the operation of the market. The state cannot be conceived of as a separate entity to the market, impotently looking on while the latter goes through its wild fluctuations of fortune, no matter how these affect national economic well-being, but rather as a player in the market, representing the interests of the entire country, or of the country's stakeholders if you will. Market failures affect far more than just investors, especially if they are on the grand scale; the Wall Street Crash was a disaster for humanity as a whole, not just those holding shares at the time. That larger context is one that the Friedmanites, with their narrowly economistic vision of the world, and *homo economicus* as their ideal, do not always acknowledge.

This should not be construed as an argument for an authoritarian state on the Chinese model, however, but for a pluralist one which accepts that the state works best when it is in close partnership with a range of non-governmental institutions as well as the market.[8] Even China may come to see the virtue of this: that the state should be regarded as the last resort, not the only resort. A properly run mixed economy should be able to attain the best of both worlds: entrepreneurial activity, but also a safety net for the economically weak. Neither the far left nor the far right will be happy to go down that route, but it is the only one that offers a system of checks and balances against a ruthless corporate sector on the one hand and an inhibiting public one on the other. A mixed economy becomes a means of negotiating the differend between these two ideologies

– and the precise balance of the mix can always be altered to suit circumstances, as well as national traditions. The issue at stake has been neatly summed up by the authors of *Beyond the Limits* (the follow-up to *The Limits of Growth*):

> *Not*: The market system will automatically bring us the future we want.
> *But*: How do we use the market system, along with many other organizational devices, to bring us the future we want?[9]

The mixed economy survived in some form even through the heyday of Friedmanite economics in the West, but only at the expense of constant criticism from the market fundamentalists, who did their very best to turn the public against the notion in their quest for 'small' government. Small government, however, is patently inadequate for the current situation.

We have seen several very high-profile examples in recent years of the state deciding that it was obliged to step in and take responsibility for market failures, to be in effect the last resort, thus giving the lie to the market fundamentalist line that public intervention can only be to the market's detriment. In the UK, the collapse of the Northern Rock bank was potentially so dangerous to the stability of the entire banking system in the country that the government had to bail out the bank and guarantee the savings of its customers – an act which required more than £20 billion plus to back it up. The alternative, as most of the banking community agreed, was probably meltdown in the British economy (various other banks suffered a run on their holdings when Northern Rock's situation became public, just to confirm the volatility of the situation). When no buyer could be found to take the responsibility off the government's hands (not much CSR being exhibited there), the company was nationalized, and although it is expected to pay back the funds that it has borrowed from the taxpayer, this will take several years – at least. As the credit crunch intensified, the British government found itself forced to become ever more active in the financial sector, orchestrating the merger of some of the country's largest banks to prevent their failure.

In the USA, the Federal Reserve became involved in similar fashion with the collapsed investment bank, Bear Sterns, and helped to negotiate an eventual buyer for it. This deal was at a knock-down price which by no means pleased the bank's shareholders, who went so far as to claim that their human rights were being abused by the Reserve's action (Northern Rock shareholders made it clear they felt just as hard done by). Once again, however, the official decision was that the risk to the system was too great for the solution to be left to the private sector alone; this was a crisis where government and its agencies could not stand idly by, where small government would have been a liability. An interesting test case for this policy soon followed when another major investment bank, Lehmann Brothers, ran into severe difficulties concerning its liquidity. This time around, a hands-off approach was adopted and the institution was allowed to go under, but its demise created such a sense of crisis that the American government felt compelled to shore up a series of other organizations who came under threat in the aftermath.

Market fundamentalist wisdom is that companies must be allowed to fail if they are badly managed, and that it distorts the market to interfere in its natural workings – particularly when it is the state that does so. If intervention is known to be forthcoming in a crisis, so the argument goes, then companies will be inclined to take risks they would not otherwise do if bankruptcy was a real threat. From a Friedmanite perspective, that makes collapses of the Northern Rock and Bear Sterns kind all the more likely to occur in future; the market has taken due note of what it would consider to be a governmental failure of nerve, and what this might license them to do. Yet not to intervene in these particular cases was to risk a stock market collapse on the scale of 1929, and neither the UK nor the American government was prepared to allow that to happen just to maintain the ideological purity of the market fundamentalist cause (and it must always be remembered that the aftermath of the Wall Street Crash was massive government intervention in the economy in the form of the Roosevelt administration's New Deal). After several decades of very free market economics, the public interest was at last deemed to take precedence. Lehmann Brothers was an experiment that is not likely to be repeated.

Since these events the market fundamentalist ethos has lost a great deal of its credibility. It is unlikely to be indulged in the West – in the immediate future, anyway, one would imagine – as it was in the heady days of the 1980s and 1990s when governments like those in the UK and the USA more or less let it run as it wanted, and, at least in principle, accepted the small government line. A lesson has been learned, and it is that public control is a necessary part of the equation for the economic system of even the most developed of nations. Liberalization and privatization cannot be seen as the solutions to all economic problems; they have their limits and these have to be marked out clearly. The market is a form of gambling (and the ultra-free market is 'casino capitalism', as one commentator back in the 1980s uncompromisingly described it[10]), and no nation should have to depend on that alone for its future. Iceland, where the financial sector has been virtually wiped out, leaving the nation bankrupt, is now bitterly regretting having done so. No matter what Milton Friedman may have claimed, extreme economic libertarianism is ultimately incompatible with the democratic ethos, which calls for a complex system of checks and balances to keep it operating smoothly and with a sense of fairness and justice to all. There is more to life than economics, and particularly more to it than abstract economic theory.

Arguments are still being trotted out that we require this economic libertarianism to inspire the entrepreneurial spirit, even after the international economic crisis of 2007–8 brought on by irresponsible bank lending in the American sub-prime mortgage market:[11] a situation which looks set to rumble on throughout the global economic system for some time to come. But we really must move past the notion that entrepreneurialism in itself is an unqualified social good, rather than an activity which has to be very carefully monitored in terms of its effect on both the public at large and the environment. The sub-prime mortgage fiasco alone is testament to the need for much tighter regulation of market activity than has existed of late in most Western economies, and governments do seem to have taken this on board. Entrepreneurialism cannot be divorced from the social structures that hold cultures together by promoting respect for the rights of others; otherwise it is all too likely to descend to the law of the jungle as its operating principle, exploiting when and where it can quite shamelessly. There are too many mavericks in the business world to take the chance that complete freedom in the market will not be abused.

One might see a socialist, or at least social democratic, lesson in all of this: that the interests of stakeholders are in the final analysis more important than those of shareholders, and that it is the legitimate business of the state to ensure that the former are given protection from the excesses of the latter. Majority rules must apply; we are all stakeholders, but we are not all shareholders. Unfortunately for the developing nations, however, globalization in its current form operates largely outside these parameters in a system where, as we saw Zygmunt Bauman complaining, 'no one now seems to be in control' (except in the sense that the free marketeers have been given largely the conditions they want). That can never be a satisfactory situation when the economic fate of entire nations is at stake. The fact that the WTO has been so unsuccessful in establishing agreements about tariffs amongst its member states in recent years (the Doha round of talks broke down in 2006, for example) is only too symbolic. Globalization, as I suggested earlier, is more like the state of nature, or the law of the jungle, than it is a democratic system, radical or otherwise. Nevertheless, we have the ability to transform it into something far more sensitive to the needs of the developing countries, into a system more about risk-sharing than exploiting the vulnerable, and we are surely under an obligation to strive to do so. The Friedmanites keep harping on about freedom, but the freedom to make tragic choices is not enough.

Where Now?

It is one thing to recommend the adoption of radical democracy and cosmopolitanism to break through the narrow nationalism and corporate greed that is serving to hold up progress on the climate change front, quite another to bring it about; but the case for them seems very strong if we are to revitalize our political narratives as I am arguing is necessary. We also need to include a healthy dose of scepticism in our revitalized narratives, and the scientific evidence for global warming has to be subjected to continuing, and searching, scrutiny, especially when the models contradict each other – no uncommon occurrence, as we have seen.[12] The best of scientific practice would guarantee this, anyway; despite what the scam-mongerers may say, scientists are in general pretty good at questioning their own beliefs. But this is not to turn us into global warming sceptics: we should be particularly sceptical about that group and the agenda they are pursuing – while acknowledging that they can still be a legitimate part of the general debate on climate change. Radical democracy would insist that they have a voice, but also that they respect the opinions of those arrayed against them. Claiming a 'scam' is not in the spirit of proper debate, but questioning the models and their projections most certainly is. Having said that, modelling remains the most effective method we have of making projections, and it is in everybody's interests that they keep being refined and studied closely.

Cosmopolitan radical democracy ought to make us all more aware of our common responsibility to bring global warming under control, and to work to keep it contained on into the future. This will require a range of actions at both a personal and collective level – reiterating the point made earlier that contributions at the former level are as much about imbuing us with the right spirit as anything else. But the critical point is that these must always be undertaken with the common good in mind, never to protect the narrow self-interest of a particular state or interest group (such as the corporate sector collectively). Due regard must always be given to the consequences, and we have to see ourselves as engaged in a global risk-sharing activity that puts us all under an obligation to refrain from taking advantage of others. Abuse of our power would backfire on us, anyway – the Earth system does not discriminate in this respect – and we really do need to have that in the forefront of our minds. We must always remember that the Earth system is very adept at recognizing false trade-offs.

It seems clear that there will have to be a range of measures to cope with climate change, and that a certain amount of experimentation will have to be done. Renewables have to be pursued; low-emission fuels must be developed further; at least in the short term, nuclear power production probably has to be expanded; globally operational legislation about carbon emissions needs to be put in place, and adherence to it closely monitored by some supranational body or bodies; public transport has to become more of a priority, particularly in the developed nations with the greatest incidence of car use; globalization as a system should be modified to be more advantageous to developing nations; the free market should be curbed in the name of social justice, and the business community required to embrace CSR as standard practice; risk-sharing amongst all the world's nations has to become the norm; we should strive to keep the world's population down as much as possible, combatting any religious objections that arise; geoengineering models must be worked on by the scientific community, continually refined, and held in reserve for worst-case scenarios (but let us pray that we manage to prevent these occurring; such schemes really have to be seen as a desperate last resort). But none of these activities will be truly effective unless there is a global change of consciousness about our relationship, and our responsibility, to the environment – and this is why green education must become a priority. We cannot go on abusing the environment as we have been doing in the recent past, and have to start regarding it as something other than an entity whose sole reason for existence is to be exploited for human material benefit, a mere 'profit center'. Trade-offs will only work if we are all involved (not just the scientists and economists), and all committed. As Ann Finlayson, a commissioner at the Sustainable Development Corporation, has emphasized, in sentiments I heartily endorse,

> We've got to start providing people, through education, with the competencies to help them understand that we have choices, and that choices have implications. Only then can we get away from the situation we're in now, in which we lurch from crisis to crisis on the environmental front.[13]

Reducing carbon emissions, pulling back from ecocide and climaticide, while guaranteeing geopolitical stability, a thriving and economically improving developing world, and all the cultural difference and diversity that goes with that, will require a considerable effort of the political imagination. Let us hope that proves to be forthcoming, that the carbon footprints wars can be resolved; the alternative is just too horrific to contemplate, with the Earth system exacting its revenge for the systematic mistreatment it has suffered at our hands since the advent of industrialization. It really is time to take out fire insurance on the planet.

Postscript

President George W. Bush's reluctance to take action to reduce his country's carbon emissions has been notorious, but both parties in the 2008 presidential election had to address the issue and they came up with very different policies. Given that America is such a massive polluter, as well as the world's richest economy, the result of that election has been a matter of considerable concern for the rest of the globe. We watched bemused as the Republican candidate, John McCain, chose to carry on much in the tradition of Bush and campaigned openly to scrap the moratorium on offshore drilling for oil, signally failing to recognize the dangers in continuing to base his country's energy policy on the use of fossil fuels. This was all the more disappointing in that McCain had previously shown some support when he was a senator for federal action to cut carbon emissions. But the election has just been won by the Democrat candidate, Barack Obama, who has an ambitious programme designed to put America in the forefront of the fight against global warming, the plan being to effect an 80 per cent reduction in the country's emissions by 2050. Amongst his goals is that 10 per cent of electricity come from renewables by 2012, and that a large-scale carbon allowance trading programme be instituted. The money the government would gain from auctioning such allowances would be spent on developing green technologies.

The elect President's policies sound promising, and demonstrate a welcome awareness of the gravity of the situation we are collectively facing. But already there are warning signs on the horizon; the banking crisis that has engulfed the world over the last few months has sparked a recession, and it is still too early to say how serious, or long-term, this will be. Action on global warming may well be put to one side as politicians concentrate on regenerating their national economies instead. Hard economic times are rarely the best contexts for innovation or large-scale cultural reorientation; the tendency instead is for societies to cling to what they know, and that may mean keeping faith with fossil fuels – as the Republicans decided. We have to be on guard to ensure that politicians like the new President actually carry out their promises, and do not allow economic crisis to take precedence over the far greater threat posed by climate change. A failing economy is undeniably a severe social problem, but a failing ecosystem is a disaster of an altogether different order.

Notes

Preface

1 See US National Oceanic and Atmospheric Administration, 'Global warming: frequently asked questions', http://www.ncdc.noaa.gov/oa/climate/globalwarming/html (accessed 17 July 2008).
2 'Ecocide' stands for 'unintended ecological suicide' (Jared Diamond, *Collapse: How Societies Choose to Fail or Survive*, London: Allen Lane, 2005, p. 6).
3 Al Gore, *An Inconvenient Truth: The Movie*, Paramount DVD, 2006.
4 Nicholas Stern, *The Economics of Climate Change: The Stern Review*, Cambridge: Cambridge University Press, 2007.

1 Introduction: The Carbon Footprint Wars: What is at Stake?

1 Brian Fagan, *The Little Ice Age: How Climate Made History 1300–1850*, New York: Basic, 2000, p. 214.
2 See, for example, his *The Revenge of Gaia: Earth's Climate Crisis and the Fate of Humanity*, London: Allen Lane, 2006.
3 James Lovelock, 'The Earth is about to catch a morbid fever that may last as long as 100,000 years', *The Independent*, 16 January, 2006, http://www.independent.co.uk/opinion/commentators/james-lovelock-the-earth-is-about-to-catch-a-morbid-fever-that-may-last-as-long-as-100000-years-523161.html (accessed 17 September 2008).

4 George Monbiot, *Heat: How We Can Stop the Planet Burning*, London: Penguin, 2007.

5 Bill McKibben, *The End of Nature: Humanity, Climate Change and the Natural World*, 2nd edn, London: Bloomsbury, 2003, p. x.

6 See Jean-François Lyotard, *The Differend: Phrases in Dispute*, trans. Georges Van Den Abbeele, Manchester: Manchester University Press, 1988. In Lyotard's more technical language, the disputants belong to discrete 'phrase regimes'.

7 A certain amount of flexibility was extended to individual countries and political units by the Treaty, with the EU being granted an 8 per cent reduction on the average, and Iceland even being allowed a 10 per cent increase.

8 Quoted in Will Hutton, *The Writing on the Wall: China and the West in the 21st Century*, 2nd edn, London: Abacus, 2008, p. 11.

9 According to recent surveys, approaching half of the world's inhabitants live in coastal areas and that percentage is steadily rising (see 'The world's coasts under threat', *New Scientist*, 1 September 2007, p. 10). Any really significant rise in sea levels would therefore have devastating consequences.

10 See, for example, Douglas Fox, 'Saved by the trees?', *New Scientist*, 20 October 2007, pp. 42–6.

11 Zygmunt Bauman, *Intimations of Postmodernity*, London: Routledge, 1992, p. 179.

12 See George Monbiot, 'Are you paying to burn the rainforest?', *The Guardian*, 4 November 2007, p. 42.

13 See Stephen A. Marglin, *The Dismal Science: How Thinking Like an Economist Undermines Community*, Cambridge, MA: Harvard University Press, 2008, Chapter 12. A tragic choice is defined as one 'where there is no good outcome, where the best outcome is the least of evils' (p. 223).

14 As Fred Pearce has pointed out in an article on the organization (whose plan is entitled 'Reducing Emissions from Deforestation and Degradation' (REDD)), much logging in developing countries is illegal and it is all too often the case, as in Papua New Guinea, that 'politicians are complicit in the illegality and profiting from it' (Fred Pearce, 'Saved?', *New Scientist*, 22 March 2008, pp. 36–9 (p. 39)).

15 Dan Smith and Janani Vivekananda, *A Climate of Conflict: The Links Between Climate Change, Peace and War*, London: International Alert, 2007, p. 3.
16 Ibid., p. 7.
17 Ibid., p. 12.
18 For Naomi Klein's analysis of the downside of the globalization ethos, see *No Logo*, London: HarperCollins, 2001, *Fences and Windows: Dispatches from the Front Lines of the Globalization Debate*, London: Flamingo, 2002, and *The Shock Doctrine: The Rise of Disaster Capitalism*, London: Penguin, 2007.
19 See in particular the work of Ernesto Laclau and Chantal Mouffe, such as their *Hegemony and Socialist Strategy: Towards a Radical Democratic Politics*, London: Verso, 1985.
20 Chantal Mouffe, *The Democratic Paradox*, London: Verso, 2000, p. 111.
21 Ulrich Beck, *World Risk Society*, Cambridge, MA: Polity Press, 1999, p. 15.
22 Bjorn Lomborg, *The Skeptical Environmentalist: Measuring the Real State of the World*, Cambridge: Cambridge University Press, 2001.
23 Fred Pearce and Bjorn Lomborg, 'An inconvenient voice' (interview with Bjorn Lomborg), *New Scientist*, 27 October 2007, pp. 54–5 (p. 55).
24 Ibid.
25 Ibid., p. 54.
26 Ibid.
27 Nicholas Stern, *The Economics of Climate Change: The Stern Review*, Cambridge: Cambridge University Press, 2007, pp. ix, xvi.

2 *Global Warming: The Evidence For*

1 See Bill McKibben, *The End of Nature: Humanity, Climate Change and the Natural World*, 2nd edn, London: Bloomsbury, 2003, p. 22.
2 Fred Pearce, *The Last Generation: How Nature Will Take Her Revenge for Climate Change*, London: Transworld, 2006, p. 340.
3 See Fred Pearce, 'We need better forecasts – and fast', *New Scientist*, 3 May 2008, pp. 8–9. The failings of modelling in terms of regional forecasting are discussed in T. N. Palmer et al., 'Toward seamless prediction: calibration of climate change projections using seasonal forecasts', *Bulletin of the American Meteorological Society* (89:4), April 2008, pp. 459–70.

4 William Laurance, 'Expect the unexpected', *New Scientist*, 12 April 2008, p. 17.

5 Pearce, *Last Generation*, p. 347.

6 Even the stock market can provide a source of such visions, with the threat of economic collapse in the very short term being a recurrent theme in recent times. The popular appetite for dystopian visions is satirized in Simon Briscoe and Hugh Aldersley-Williams, *Panicology*, London: Penguin, 2008, where the authors ponder such issues as why we are so worried by bird flu (no deaths from which as yet recorded in the UK), when 12,000 people a year die from ordinary flu in the country.

7 Alison Benjamin and Brian McCallum, *A World Without Bees*, London: Guardian Books, 2008.

8 'For whom the bell tolls', *New Scientist*, 5 April 2008, p. 5; Donella H. Meadows et al., *The Limits to Growth: A Report for the Club of Rome's Project on the Predicament of Mankind*, New York: Potomac Associates, 1972.

9 Donella H. Meadows, Dennis L. Meadows, and Jorgen Randers, *Beyond the Limits: Global Collapse or a Sustainable Future*, London: Earthscan, 1992, p. xiii.

10 The feature consisted of two articles by Debora MacKenzie, 'The end of civilization' and 'Are we doomed?' (*New Scientist*, 5 April 2008, pp. 28–31 and pp. 33–5).

11 In an interview, the scientist Chris Langton described the benefits of existence at the edge of chaos as follows: 'Being at the transition point between order and chaos not only buys you exquisite control – small input/big change – but it also buys you the possibility that information processing can become an important part of the dynamics of the system' (Roger Lewin, *Complexity: Life on the Edge of Chaos*, London: Phoenix, 1993, p. 51).

12 Debora MacKenzie, 'The end of civilization', p. 35. The team was located at Los Alamos National Laboratory, New Mexico.

13 James Lovelock, *The Revenge of Gaia: Why the Earth is Fighting Back – and How We Can Still Save Humanity*, London: Penguin, 2007, p. 7.

14 Pearce, *Last Generation*, p. 350.

15 The article was referring to the fourth Global Environment Outlook (GEO) report, published by the UN Environment Programme (2007), which came to the conclusion that 'we need 1.4 earths' to cope with

the current population of 6.7 billion ('We'd like another half a planet please', *New Scientist*, 3 November 2007, p. 13). Given that the world's population continues to increase at an alarming rate (predicted to reach between 8 and 10 billion by 2050), this is dispiriting news.

16 Meadows et al., *Beyond the Limits*, p. 2.

17 Pearce, *Last Generation*, p. 349.

18 Mark Lynas, *Six Degrees: Our Future on a Hotter Planet*, London: HarperCollins, 2007, pp. 214, 223.

19 On the Permian period, see Michael J. Benton, *When Life Nearly Died: The Greatest Mass Extinction of All Time*, London: Thames & Hudson, 2003. Benton emphasizes that the precise causes of the extinction event are still unclear, but that the evidence points to global warming as the most critical factor in accelerating the process.

20 Dan McDougall, 'Time runs out for islanders on global warming's front line', *The Guardian*, 30 March 2008, p. 46.

21 See McKibben, *End of Nature*, pp. 150–1.

22 See Gabrielle Walker, *Snowball Earth: The Story of the Great Global Catastrophe That Spawned Life As We Know It*, London: Bloomsbury, 2003.

23 George Monbiot, *Heat: How We Can Stop the Planet Burning*, 2nd edn, London: Penguin, 2007, pp. xiv–xv.

24 Ibid., p. 89.

25 Ibid., p. 91.

26 Gabrielle Walker and Sir David King, *The Hot Topic: How to Tackle Global Warming and Still Keep the Lights On*, London: Bloomsbury, 2008, p. 148.

27 Jared Diamond, *Collapse: How Societies Choose to Fail or Survive*, London: Allen Lane, 2005, p. 504.

28 Walker and King, *Hot Topic*, p. 179.

29 Ibid., p. 255.

30 McKibben, *End of Nature*, p. xiv.

31 Ibid., p. xviii.

32 Ibid., p. 210.

33 Bill McKibben, *Hope, Human and Wild: True Stories of Living Lightly on the Earth*, Little, Brown: London, 1995.

34 Nicholas Stern, *The Economics of Climate Change: The Stern Review*, Cambridge: Cambridge University Press, 2007, p. xiv.

35 Quoted in Pearce, *Last Generation*, p. 87.

36 McKibben, *End of Nature*, p. ix.
37 Quoted in Pearce, 'We need better forecasts', p. 8.
38 Sarah B. Das et al., 'Fracture propagation to the base of the Greenland ice sheet during supraglacial lake drainage', *Science* (320: 5877), 9 May 2008, pp. 778–81 [DOI: 10.1126/science.1153360].
39 McKibben, *End of Nature*, p. 108.
40 Quoted in Ed Pilkington, 'Climate target is guaranteed catastrophe', *The Guardian*, 7 April 2008, p. 1.
41 Quoted in ibid.
42 In Fred Pearce's graphic description, life on Earth was devastated by a 'megafart', and the potential for another is still there in the form of methane clathrates trapped under the oceans (*The Last Generation*, p. 150). It has been estimated that anywhere between 1 and 10 trillion tonnes of methane is currently locked up in these clathrates, and they require the oceans to stay cold to remain that way. If the oceans warm up, as they would with any pronounced global warming, then the situation would of course change.
43 Will Hutton, *The Writing on the Wall: China and the West in the 21st Century*, 2nd edn, London: Abacus, 2008, p. 178.

3 Global Warming: The Arguments Against

1 Joseph Tainter, quoted in Debora MacKenzie, 'Are we doomed?', *New Scientist*, 5 April 2008, pp. 32–5 (p. 35).
2 Jared Diamond, *Collapse: How Societies Choose to Fail or Survive*, London: Allen Lane, 2005, p. 15.
3 Bjorn Lomborg, *The Skeptical Environmentalist: Measuring the Real State of the World*, Cambridge: Cambridge University Press, 2001.
4 See Brian Fagan, *The Little Ice Age: How Climate Made History 1300–1850*, New York: Basic, 2000. As Fagan notes, the dates are only approximate, and for some climate historians the Little Ice Age is a late seventeenth- to mid-nineteenth-century phenomenon instead.
5 Ibid., p. 21. Wine growing in England ceased to be viable by the mid-fifteenth century.
6 Reported on in Fred Pearce, 'Rising temperatures bring their own CO_2', *New Scientist*, 22 March 2008, p. 11. As one of the study's team

put it, '[p]eople on both sides want a one-way link, but the historical record shows that causality goes both ways . . . Actually, CO_2 is more sensitive to temperature than the other way round' (quoted in ibid.). The problem with this analysis is that it envisages carbon emissions increasing naturally on top of those created by humankind, making the task of reducing them considerably more complicated for all concerned (see Martin Scheffer, Viktor Brovkin, and Peter M. Cox, 'Positive feedback between global warming and atmospheric CO_2 concentration', *Geophysical Research Letters* (33), 26 May 2006, L10702).

7 Brian Fagan, *The Great Warming: Climate Change and the Rise and Fall of Civilizations*, New York: Bloomsbury, 2008; Diamond, *Collapse*. Also see Joseph A. Tainter, *The Collapse of Complex Societies*, Cambridge: Cambridge University Press, 1988.

8 Fagan, *The Great Warming*, p. 229.

9 Tainter, *Collapse of Complex Societies*, p. 216.

10 Fagan, *Little Ice Age*, p. 123.

11 Quoted in David Sington, 'Global dimming', http://www.bbc.co.uk/sn/tvradio/programmes/horizon/dimming_prog_summary.shtml (accessed 25 April 2008).

12 Quoted in Michael Schirber, 'Ice ages blamed on tilted Earth', http://www.livescience.com/environment/050330_earth_tilt.html, 30 March 2008 (accessed 8 May 2008).

13 See, for example, James G. Titus et al., 'Greenhouse effect and sea level rise: the cost of holding back the sea', http://www.epa.gov/climatechange/effects/downloads/cost_of_holding.pdf, p. 5.

14 Bill McKibben, *The End of Nature: Humanity, Climate Change and the Natural World*, 2nd edn, London: Bloomsbury, 2003, p. 126.

15 See N. S. Keenlyside et al., 'Advancing decadal-scale climate prediction in the North Atlantic sector', *Nature* (453), 1 May 2008, pp. 84–7.

16 'Warming oceans starved of oxygen', *New Scientist*, 10 May 2008, p. 17.

17 See US National Oceanic and Atmospheric Administration, 'Global warming: frequently asked questions', http://www.ncdc.noaa.gov/oa/climate/globalwarming/html (accessed 17 July 2008).

18 Douglas Fox, 'Saved by the trees?', *New Scientist*, 20 October 2007, pp. 42–6 (p. 42).

19 Ibid., p. 46.

20 See Fred Pearce, 'Making a difference', *New Scientist*, 12 April 2008, pp. 50–3. Climate Care also ran a scheme in 2007 whereby the public could offset their holiday flights, thus contributing towards a wind turbine project the group were involved in setting up in China (see Fred Pearce, 'Dirty, sexy money', *New Scientist*, 19 April 2008, pp. 38–41 (p. 39)).
21 Pearce, 'Making a difference', p. 53.
22 Fred Pearce, *Confessions of an Eco-Sinner: Travels to Find Where My Stuff Comes From*, London: Eden Project, 2008, p. 304.
23 Tom Burke, quoted in Pearce, 'Dirty, sexy money', p. 38.
24 Quoted in ibid.
25 Lomborg, *Skeptical Environmentalist*, p. xx.
26 Ibid., p. 4.
27 Ibid., p. 7.
28 Bjorn Lomborg, *Cool It: The Skeptical Environmentalist's Guide to Global Warming*, New York: Knopf, 2007, pp. 5, 6.
29 Ibid., p. 6.
30 Ibid.
31 Al Gore, *An Inconvenient Truth: The Movie*, Paramount DVD, 2006.
32 Lomborg, *Cool It*, p. 226.
33 J. E. Hansen, 'A slippery slope: how much global warming constitutes "dangerous anthropogenic interference"?', *Climatic Change* (68:3), 2005, pp. 269–79 (p. 278).
34 Lomborg, *Cool It*, p. 199.
35 Ibid., p. 201.
36 Nigel Lawson, *An Appeal to Unreason: A Cool Look at Global Warming*, London: Duckworth Overlook, 2008, pp. 11–12, 21.
37 Wallace S. Broecker and Robert Kunzig, *Fixing Climate: What Past Climate Changes Reveal About the Current Threat – and How to Counter It*, New York: Hill & Wang, 2008, p. xii.
38 Lawson, *Appeal*, p. 106.
39 Richard Lambert, 'Fuelling the debate on climate change', *The Guardian*, Review Section, 19 April 2008, p. 9.
40 Lawson, *Appeal*, p. 12.
41 Ibid., p. 101.
42 Essentially the line also taken by Fred Krupp and Miriam Horn in *Earth: The Sequel. The Race to Reinvent Energy and Stop Global Warming*, New

York: W. W. Norton, 2008: 'We have before us an extraordinary opportunity: to harness the power of the United States of America's huge and dynamic markets to ensure a safe future' (p. 252)

43 Richard Lindzen, 'Climate of fear: Global-warming alarmists intimidate dissenting scientists into silence', *Wall Street Journal*, 12 April 2006, http://www.opinionjournal.com/extra/?id=110008220 (accessed 16 July 2008).

44 Quoted in Fred Pearce, 'Climate change: menace or myth?', *New Scientist*, 12 February 2005, pp. 38–43 (p. 40).

45 http://www.iceagenow.com and http://www.JunkScience.com, for example.

46 Fagan, *Little Ice Age*, p. 59.

47 Diamond, *Collapse*, p. 11.

48 Fagan, *Little Ice Age*, p. 101.

4 The Globalization Paradigm: Defenders and Detractors

1 See David Stuckler, Lawrence P. King, and Sanjay Basu, 'International Monetary Fund Programs and Tuberculosis Outcomes in Post-Communist Countries', *PLoS Medicine* (5:7), July 2008, DOI: 10.1371/journal.pmed.0050143. The study reveals a marked increase in TB cases in Russia and Eastern Europe as public spending was slashed to meet IMF targets.

2 Joseph Stiglitz, *Globalization and its Discontents*, London: Penguin, 2002, p. 130.

3 Ibid., p. 131.

4 Will Hutton, *The Writing on the Wall: China and the West in the 21st Century*, 2nd edn, London: Abacus, 2008, p. 14.

5 Joseph Stiglitz, *Making Globalization Work*, London: Penguin, 2006, p. 285.

6 Ibid., p. 101.

7 Ibid., p. 292.

8 Naomi Klein, *No Logo*, London: HarperCollins, 2001, p. xvii.

9 Ibid., p. 446.

10 Naomi Klein, *The Shock Doctrine: The Rise of Disaster Capitalism*, London: Allen Lane, 2007, p. 49.

11 Naomi Klein, *Fences and Windows: Dispatches from the Front Lines of the Globalization Debate*, London: Flamingo, 2002, p. xv.
12 Zygmunt Bauman, *Intimations of Postmodernity*, London: Routledge, 1992, p. 175.
13 Zygmunt Bauman, *Globalization: The Human Consequences*, Cambridge: Polity, 1998, p. 123.
14 Ibid., p. 58.
15 Ibid., p. 112.
16 Rachel Louise Snyder, *Fugitive Denim: A Moving Story of People and Pants in the Borderless World of Global Trade*, New York: W. W. Norton, 2008, p. 316.
17 Ibid., p. 21.
18 Fred Pearce, *Confessions of an Eco-Sinner: Travels to Find Where My Stuff Comes From*, London: Eden Project, 2008, p. 4.
19 Ibid., p. 10.
20 Martin Wolf, *Why Globalization Works*, New Haven, CT: Yale University Press, 2004, p. 95. For another defence of globalization as a means of eradicating poverty, see Donald J. Boudreaux, *Globalization*, Westport, CT: Greenwood, 2007.
21 Wolf, *Why Globalization Works*, p. 4.
22 Ibid. Sylvia Ostry had earlier dismissed the anti-capitalist movement as 'dissent.com' (see Wolf, ibid.).
23 Ibid., p. 8.
24 Ibid., p. 12.
25 Ibid., p. 172.
26 Milton Friedman, *Capitalism and Freedom*, Chicago: University of Chicago Press, 1982, p. vi.
27 Ibid., p. 3.
28 Ibid.
29 Ibid., p. 25.
30 Ibid., p. 9.
31 Ibid., p. 67.
32 Stephen A. Marglin, *The Dismal Science: How Thinking Like an Economist Undermines Community*, Cambridge, MA: Harvard University Press, 2008, p. 3.
33 Ibid., p. x.
34 Ibid., p. 56.

35 Ibid., p. 50.

36 Terry Macalister, 'A change in the climate: credit crunch makes the bottom line the top issue', *The Guardian*, 6 March 2008, p. 28.

37 Ibid.

38 Ibid.

39 Quoted in Terry Macalister, '"Dishonest, irresponsible": Shell lambasted for pulling out of world's biggest wind farm', *The Guardian*, 2 May 2008, p. 36.

40 Jared Diamond, *Collapse: How Societies Choose to Fail or Survive*, London: Allen Lane, 2005, p. 38.

41 Ibid., p. 38.

42 John Gray, 'Those who control oil and water will control the world', *The Guardian*, 30 March 2008, p. 33.

43 Hutton, *The Writing on the Wall*, pp. xii, xi.

5 Reducing Our Carbon Footprint: Altering Lifestyles

1 Nicholas Stern, *The Economics of Climate Change: The Stern Review*, Cambridge: Cambridge University Press, 2007, p. xiii.

2 Ibid., p. 327.

3 Ibid., p. 654.

4 As Brian Fagan points out in *The Little Ice Age: How Climate Made History 1300–1850* (New York: Basic Books, 2000), '[e]xcavations in medieval cemeteries paint a horrifying picture of health problems resulting from brutal work regimes,' such as spinal deformations, arthritis, and osteoarthritis from the unremittingly hard manual labour required by farming and fishing. Average life-spans were, as well, radically lower than today – down into the twenties in most cases.

5 Paul R. Ehrlich, *The Population Bomb*, New York: Ballantine, 1968. Ehrlich's pessimistic assessment of our prospects was strongly countered by such as Julian L. Simon (see, for example, *The Ultimate Resource*, Princeton, NJ; Princeton University Press, 1981).

6 Bill Devall and George Sessions, *Deep Ecology: Living as if Nature Mattered*, Salt Lake City, UT: Peregrine Smith, 1985, pp. 7, 9.

7 See George Monbiot, *Heat: How We Can Stop the Planet Burning*, 2nd edn, London: Penguin, 2007, p. xxii.

8 George Monbiot, 'If there is a God, he's not green. Otherwise airships would take off', *The Guardian*, 6 May 2008, p. 25.
9 Stern, *Stern Review*, p. 389.
10 See Royal Commission on Environmental Pollution, 'The environmental effects of civil aircraft in flight', http://www.rcep.org.uk/aviation/av12-txt.pdf, p. 20 (accessed 17 July 2008).
11 Monbiot, 'If there is a God'.
12 Lynn Sloman, *Car Sick: Solutions for Our Car-addicted Culture*, Totnes: Green Books, 2006. The author advocates systematic 'de-motorisation' of our cities.
13 Several studies in the 1980s showed an apparent increase in childhood leukaemia near nuclear power stations in the UK, but the government's advisers rejected the idea of there being a causal connection. More recent studies in Europe and America have reactivated the debate, however, in some cases claiming rates of childhood leukaemia of 14–21 per cent higher near nuclear plants (for a survey of those studies, see Ian Fairlie, 'Reasonable doubt', *New Scientist*, 26 April 2008, p. 18).
14 Terry Macalister, 'Dawn of a new nuclear age', *The Guardian*, 22 March 2008, p. 47.
15 William Laurance, 'Expect the unexpected', *New Scientist*, 12 April 2008, p. 17.
16 Quoted in Macalister, 'Dawn of', p. 47.
17 John Large, quoted in ibid., p. 47.
18 Quoted in Andrew Sparrow and Patrick Wintour, 'Nuclear is UK's new North Sea oil - minister', *The Guardian*, 26 March 2008, p. 1.
19 William Calvin, *Global Fever: How to Treat Climate Change*, Chicago: University of Chicago Press, 2008, p. 9.
20 Brian Clegg, *The Global Warming Survival Kit: The Must-Have Guide to Overcoming Extreme Weather, Power Cuts, Food Shortages and Other Climate Change Disasters*, London: Doubleday, 2007 (back cover quote).
21 Ibid., p. 37.
22 Fred Pearce, 'Tracking the winds of climate change', *New Scientist*, 26 April 2008, pp. 48–9 (p. 49).

6 Living With Our Carbon Footprint: The Technological Response

1 Quoted in Terry Macalister, 'Biofuels: a solution that became part of the problem', *The Guardian*, 25 March 2008, p. 11.
2 Terry Macalister, 'Demands for crackdown on biofuels scam', *The Guardian*, 1 April 2008, p. 1.
3 'An unsustainable scam', *The Guardian*, 1 April 2008, p. 34.
4 Roger Angel, interviewed by Robyn Williams, 'Sunshade in space', The Science Show, ABC Radio National Australia, 11 November 2006, http://www.abc.net.au/rn/scienceshow/stories/2006/1785912.htm (accessed 26 May 2008).
5 Molly Bentley, 'Guns and sunshades to rescue climate', http://news.bbc.co.uk/1/hi/sci/tech/4762720.stm, 2 March 2006 (accessed 26 May 2008).
6 Quoted in 'We could use moon as giant flashlight', *New Scientist*, 26 April 2008, p. 17.
7 Richard Somerville, quoted in Bentley, 'Guns and sunshades'.
8 See Alvia Gaskill, 'Global albedo enhancement project', *Library for Science*, http://www.global-warming-geo-engineering.org/Albedo-Enhancement/Introduction/ag1.html, 1.3 (accessed 26 May 2008).
9 Ibid., 4.1.
10 Ibid., 2.4.5. There is an interesting echo in Gaskill's scheme of the artist Christo's work, such as wrapping up the Pont Neuf in Paris, and covering up a 1 mile stretch of coastline in Australia with plastic sheeting. These were both only short-term efforts, of course.
11 James Lovelock and Chris Rapley, 'Ocean pipes could help the Earth to cure itself', *Nature* (449), 27 September 2007, p. 403.
12 See Nin Zeng, 'Carbon sequestration via wood burial', *Carbon Balance and Management* (3:1), 3 January 2008 (http://cbmjournal.com/content/3/1/1).
13 Richard Lovett, 'Carbon lockdown', *New Scientist*, 3 May 2008, pp. 32–5 (p. 32).
14 Quoted in ibid., p. 35.
15 See Ken Zweibel, James Mason, and Vasilis Fthenakis, 'A grand solar plan', *Scientific American*, January 2008, pp. 64–73.
16 'Soaring commodity prices . . . have been an unanticipated boon to the

coal producing regions of countries like Japan that had written off coal mining as a relic of the Industrial Revolution' (Martin Fackler, 'As oil prices rise, nations revive coal mining', *New York Times*, 22 May 2008, topics/nytimes.com/top/reference/timestopics/people/f/martin_fackler/index.html? (accessed 16 July 2008)).

17 Monbiot is equally sceptical of UK government claims that a new generation of nuclear power stations will create 100,000 jobs, asking pointedly, 'When and how? Here, or in France?' (George Monbiot, 'Jobs are used to justify anything, but the numbers don't add up', *The Guardian*, 1 April 2008, p. 31). In similar vein, Bill McKibben has noted that, '[a]ny numbers are at best a guess, useful only as a way of saying "big problem, very big problem"'(*The End of Nature: Humanity, Climate Change and the Natural World*, 2nd edn, London; Bloomsbury, 2003, p. 122).

18 Bjorn Lomborg, *Cool It: The Skeptical Environmentalist's Guide to Global Warming*, London: Marshall Cavendish, 2007, p. 158.

19 Ibid., p. 16.

20 Quoted in Julian Glover, 'They cheat, I tell you', *The Guardian*, 3 May 2008, p. 35.

7 Worst-Case Scenarios: Economic

1 Rosie Boycott, 'Only a radical change of diet can halt looming food crises', *The Guardian*, 28 March 2008, p. 35.

2 Quoted in Ibid.

3 Tim Lang, *Food Industrialisation and Food Power: Implications for Food Governance*, London: IIED, 2004, p. 17.

4 'Overview of the benefits and carbon costs of the African horticultural trade with the UK', Food Chain Economics Unit, DEFRA, http://www.dfid.gov.uk/news/files/speeches/trade/hilary-roses-feb07.asp (accessed 12 May 2008).

5 Fred Pearce, *Confessions of an Eco-Sinner: Travels to Find Where My Stuff Comes From*, London: Eden Project, 2008, p. 111.

6 Jared Diamond, *Collapse: How Societies Choose to Fail or Survive*, London: Allen Lane, 2005, p. 159.

7 José Montilla, quoted in Graham Keeley, 'Drought ignites Spain's "water war"', *The Observer*, 6 April 2008, p. 37.

8 Dan McDougall, 'Time runs out for islanders on global warming's front line', *The Guardian*, 30 March 2008, p. 46.

9 Naomi Klein, *Fences and Windows: Dispatches from the Front Lines of the Globalization Debate*, London: Flamingo, 2002, p. xviii.

8 Worst-Case Scenarios: Socio-Political

1 World Travel and Tourism Council, http://www.wttc.org/eng/Tourism_News/Press_Releases/Press_Releases_2008/Continued_growth_signalled_for_Travel_and_Tourism_Industry/ (accessed 23 June 2008).

2 In a speech delivered at the World Summit on Sustainable Development, Johannesburg, 2002.

3 World Travel and Tourism Council, http://www.wttc.org/eng/Tourism_Research/Tourism_Satellite_Accounting/TSA_Country_Reports/Peru/ (accessed 24 June 2008).

4 'Travel and tourism in Gambia', Euromonitor International, http://www.euromonitor.com/Travel_And_Tourism_in_Gambia (accessed 11 May 2008).

5 See 'The Gambia tourism value chain and its pro-poor impact', Overseas Development Institute, http://www.odi.org.uk/tourism/projects/06_gambiavca_html (accessed 11 May, 2008).

6 Thomas Hardy, *The Return of the Native* [1878], ed. George Woodcock, London: Penguin, 1978, p. 55.

7 World Travel and Tourism Council, http://www.wttc.org/eng/Tourism_Research/Tourism_Satellite_Accounting/TSA_Country_Reports/Greece/ (accessed 23 June 2008).

8 World Travel and Tourism Council, http://www.wttc.org/eng/Tourism_Research/Tourism_Satellite_Accounting/TSA_Country_Reports/France/ (accessed 23 June 2008).

9 VisitBritain, http://www.tourismtrade.org.uk/MarketIntelligenceResearch/KeyTourismFacts.asp (accessed 6 August 2008).

10 See Fred Pearce, *Confessions of an Eco-Sinner: Travels to Find Where My Stuff Comes From*, London: Eden Project, 2008, Chapter 35.

9 Worst-Case Scenarios: Technological and Environmental

1 See Xan Rice, 'Uganda "averts tragedy" with reversal of decision to clear virgin forest for biofuel', *The Guardian*, 29 October 2007, p. 22.
2 See Xan Rice, 'Wildlife and livelihoods at risk in Kenyan wetlands biofuel project', *The Guardian*, 24 June 2008, p. 17.
3 See D. J. Lunt et al., '"Sunshade world": a fully coupled GCM evaluation of the climatic impacts of geoengineering', *Geophysical Research Letters*, (35), 25 June 2008, L12710.
4 'Sunshield to cool planet would decimate ozone layer', *New Scientist*, 3 May 2008, p. 17.
5 See, for example, Dixy Lee Ray, with Lou Guzzo, *Trashing the Planet: How Science Can Help Us Deal with Acid Rain, Depletion of the Ozone, and Nuclear Waste (Among Other Things)*, Washington, DC: Regnery, 1990. As far as the authors are concerned, 'we have been panicked into spending billions of dollars to cure problems without knowing whether they are real' (ibid., p. ix).
6 Alvia Gaskill, 'Global albedo enhancement project', *Library for Science*, http://www.global-warming-geo-engineering.org/Albedo-Enhancement/Introduction/ag1.html, 1.3 (accessed 26 May 2008).
7 For more on this see Chapter 4 of Jared Diamond, *Collapse: How Societies Choose to Fail or Survive*, London: Allen Lane, 2005.
8 Ken Zweibel, James Mason, and Vasilis Fthenakis, 'A solar grand plan', *Scientific American*, January 2008, pp. 64–73.
9 David Sassoon, 'Breakthrough: concentrated solar power all over southwest US', http://solveclimate.com/blog/20080117/breakthrough-concentrated-solar-power-all-over-southwest-us, 17 January 2008 (accessed 22 May 2008).
10 Worldwatch Institute and Center for American Progress, 'American energy: the renewable path to energy security', http://americanenergynow.net/AmericanEnergy.pdf, p. 30 (accessed 22 May 2008).
11 Windmills are another matter, and painters have been attracted by them down through the years. But one has only to think of what John Constable's 'Flatford Mill' would look like were it to become 'Flatford Wind Farm', with the landscape filled with the towers and their huge propellers, to appreciate what the aesthetic difference would be.

12 Julian Borger, 'UN warns of new face of hunger as global prices soar', *The Guardian*, 26 February 2008, p. 18.
13 John Vidal, 'The looming food crisis', *The Guardian*, G2 Section, 29 August 2007, pp. 5–9 (p. 5).
14 Cynthia Rosenzweig et al., 'Attributing physical and biological impacts to anthropogenic climate change', *Nature* (453), 15 May 2008, pp. 353–7. The survey details how plant and animal life around the globe has been affected by environmental change since 1970, concluding that as much as 90 per cent of recorded examples can be put down to the effect of global warming.
15 Eric Chivian and Aaron Bernstein, 'Threatened groups of organisms valuable to medicine', in Eric Chivian and Aaron Bernstein, eds, *Sustaining Life: How Human Health Depends on Biodiversity*, New York: Oxford University Press, 2008, pp. 203–85 (pp. 230–1).

10 Reconstructing Geopolitical Relationships: The Ethical Dimension

1 Jared Diamond, *Collapse: How Societies Choose to Fail or Survive*, London: Allen Lane, 2005, p. 134
2 Bjorn Lomborg, *The Skeptical Environmentalist: Measuring the Real State of the World*, Cambridge: Cambridge University Press, 2001, p. 29.
3 Edward Said, *Orientalism*, London: Routledge & Kegan Paul, 1978, p. 1.
4 Fred Pearce, *Confessions of an Eco-Sinner: Travels to Find Where My Stuff Comes From*, London: Eden Project, 2008, p. 78.
5 See Lawrence Wright, 'The rebellion within', *The Observer Magazine*, 13 July 2008, pp. 14–49.

11 Reconstructing Geopolitical Narratives: A Radical Democratic Globe?

1 The US steel industry has been the recipient of substantial subsidies from the federal government in the last few decades, a fact not unconnected with the impact this can have on voting in the states where it has a large presence. Others to receive similar treatment from the American government include the cotton industry, the world's largest, which is creating a crisis elsewhere, notably in its Egyptian counterpart. The

latter is finding it increasingly difficult to compete on the world market, despite the superiority of its basic product.

2 Julian Baggini, 'Philosophical globalisation', *The Guardian*, Review Section, 30 August 2008, p. 5.

3 Ernesto Laclau and Chantal Mouffe, *Hegemony and Socialist Strategy: Towards a Radical Democratic Politics*, London: Verso, 1985.

4 Chantal Mouffe, *The Democratic Paradox*, London: Verso, p. 96. For more recent perspectives on radical democracy, see Lars Tonder and Lasse Thomassen, eds, *Radical Democracy: Politics Between Abundance and Lack*, Manchester: Manchester University Press, 2005.

5 Ibid., p. 101.

6 David R. Howarth, 'Ethos, agonism and populism: William Connolly and the case for radical democracy', *British Journal of Politics and International Relations* (10:2), 2008, pp. 171–93 (p. 189).

7 Ibid.

8 Jason Glynos and David Howarth, *Logics of Critical Explanation in Social and Political Theory*, London: Routledge, 2007, p. 167.

9 Chantal Mouffe, *On the Political*, London: Routledge, 2005, p. 1.

10 Ibid., p. 5.

11 Ulrich Beck, *World Risk Society*, Cambridge, MA: Polity, 1999, p. 15.

12 Ibid., pp. 14, 15.

13 Ibid., p. 16.

14 Ibid., p. 3.

15 Ibid., p. 91.

16 Ibid., p. 8.

17 Joanna Moorhead, 'Winds of change', *The Guardian* (Supplement, 'The green mile'), 30 August 2008, p. 1.

18 Mouffe, *On the Political*, 120.

19 Danielle Archibugi, 'From the United Nations to cosmopolitan democracy', in Danielle Archibugi and David Held, eds, *Cosmopolitan Democracy: An Agenda for a New World Order*, Cambridge and Oxford: Polity and Blackwell, 1995, pp. 121–62 (p. 122).

20 See Thomas Hobbes, *Leviathan, or The Matter, Forme, and Power of a Common-Wealth Ecclesiasticall and Civill* (1651), ed. Richard Tuck, Cambridge: Cambridge University Press, 1991, Chapter 13.

21 '[T]he weakest has strength enough to kill the strongest, either by secret

machination, or by confederacy with others, that are in the same danger with himselfe' (ibid., p. 87).

12 Conclusion: Survival, Disaster, Trade-Off

1 Quoted in John Vidal, 'Government figures hid scale of CO_2 emissions, says report', *The Guardian*, 17 March 2008, p. 1. The National Audit Office was highly critical of the whole exercise of record-keeping, noting disapprovingly that government departments often drew very different conclusions from each other from the same data (see National Audit Office, 'UK greenhouse gas emissions: measurement and reporting', Spring, 2008, http://www.nao.org.uk/publications/0708_greenhouse_gas_emissions.pdf (accessed 17 July, 2008)).

2 John Sauven, quoted in John Vidal, 'Britain seeks loophole in EU green energy targets', *The Guardian*, 29 March 2008, p. 1.

3 Quoted in ibid.

4 Ulrich Beck, 'This free-market farce shows how badly we need the state', *The Guardian*, 10 April 2008, p. 35.

5 Fred Pearce, *Confessions of an Eco-Sinner: Travels to Find Where My Stuff Comes From*, London: Eden Project, 2008, p. 357.

6 Quoted in Fred Krupp and Miriam Horn, *Earth: The Sequel. The Race to Reinvent Energy and Stop Global Warming*, New York: W. W. Norton, 2008, p. 6.

7 Gabrielle Walker and Sir David King, *The Hot Topic: How to Tackle Global Warming and Still Keep the Lights On*, London: Bloomsbury, 2008, p. 178.

8 For a discussion of the balancing act this requires, see Andrew Dunsire, 'Tipping the balance: autopoeisis and governance', *Administration and Society* (28:3), 1996, pp. 299–334. The author's preference is for what he calls 'collibration', described as '[a] technique of government action built upon the encouragement and maintenance of social self-policing and self-regulation' (p. 328).

9 Donella H. Meadows, Dennis L. Meadows, and Jorgen Randers, *Beyond the Limits: Global Collapse or a Sustainable Future*, London: Earthscan, 1992, p. 230.

10 Susan Strange, *Casino Capitalism*, Oxford: Blackwell, 1986.

11 See, for example, Ruth Lea, 'Spare us the meddlers', *The Guardian*, 10 April 2008, p. 35.
12 I discuss the virtues of scepticism in social and political life in general in my *Empires of Belief: Why We Need More Scepticism and Doubt in the Twenty-First Century*, Edinburgh: Edinburgh University Press, 2006.
13 Quoted in Joanna Moorhead, 'Winds of change', *The Guardian* (Supplement, 'The green mile'), 30 August 2008, p. 1.

Bibliography

Angel, Roger and Williams, Robyn, 'Sunshade in space', The Science Show, ABC Radio National Australia, 11 November 2006, http://www.abc.net.au/rn/scienceshow/stories/2006/1785912.htm (accessed 26 May 2008).

Archibugi, Danielle, 'From the United Nations to cosmopolitan democracy', in Danielle Archibugi and David Held, eds, *Cosmopolitan Democracy: An Agenda for a New World Order*, Cambridge and Oxford: Polity Press and Blackwell, 1995, pp. 121–62.

— and Held, David, eds, *Cosmopolitan Democracy: An Agenda for a New World Order*, Cambridge and Oxford: Polity Press and Blackwell, 1995.

Baggini, Julian, 'Philosophical globalisation', *The Guardian*, Review Section, 30 August 2008, p. 5.

Bauman, Zygmunt, *Intimations of Postmodernity*, London: Routledge, 1992.

— *Globalization: The Human Consequences*, Cambridge: Polity, 1999.

Beck, Ulrich, *World Risk Society*, Cambridge, MA: Polity, 1999.

— 'This free-market farce shows how badly we need the state', *The Guardian*, 10 April 2008, p. 35.

Benjamin, Alison and McCallum, Brian, *A World Without Bees*, London: Guardian Books, 2008.

Bentley, Molly, 'Guns and sunshades to rescue climate', http://news.bbc.co.uk/1/hi/sci/tech/4762720.stm, 2 March 2006 (accessed 26 May 2008).

Benton, Michael J., *When Life Nearly Died: The Greatest Mass Extinction of All Time*, London: Thames & Hudson, 2003.

Borger, Julian, 'UN warns of new face of hunger as global prices soar', *The Guardian*, 26 February 2008, p. 18.

Boudreaux, Donald J., *Globalization*, Westport, CT: Greenwood, 2007.

Boycott, Rosie, 'Only a radical change of diet can halt looming food crises', *The Guardian*, 28 March 2008, p. 35.

Briscoe, Simon and Aldersley-Williams, Hugh, *Panicology*, London: Penguin, 2008.

Broecker, Wallace S. and Kunzig, Robert, *Fixing Climate: What Past Climate Changes Reveal About the Current Threat – and How to Counter It*, New York: Hill & Wang, 2008.

Calvin, William, *Global Fever: How to Treat Climate Change*, Chicago: University of Chicago Press, 2008.

Chivian, Eric and Bernstein, Aaron, eds, *Sustaining Life: How Human Health Depends on Biodiversity*, New York: Oxford University Press, 2008.

— 'Threatened groups of organisms valuable to medicine', in Eric Chivian and Aaron Bernstein, eds, *Sustaining Life: How Human Health Depends on Biodiversity*, New York: Oxford University Press, 2008, pp. 203–85.

Clegg, Brian, *The Global Warming Survival Kit: The Must-Have Guide to Overcoming Extreme Weather, Power Cuts, Food Shortages and Other Climate Change Disasters*, London: Doubleday, 2007.

Das, Sarah B. et. al., 'Fracture propagation to the base of the Greenland ice sheet during supraglacial lake drainage', *Science* 320 (5877), 9 May 2008, pp. 778–81 [DOI: 10.1126/science.1153360].

Defra, Food Chain Economics Unit, 'Overview of the benefits and carbon costs of the African horticultural trade with the UK', http://www.dfid.gov.uk/news/files/speeches/trade/hilary-roses-feb07.asp (accessed 12 May 2008).

Devall, Bill and Sessions, George, *Deep Ecology: Living as if Nature Mattered*, Salt Lake City, UT: Peregrine Smith, 1985.

Diamond, Jared, *Collapse: How Societies Choose to Fail or Survive*, London: Allen Lane, 2005.

Dunsire, Andrew, 'Tipping the balance: autopoeisis and governance', *Administration and Society* (28:3) 1996, pp. 299–334.

Ehrlich, Paul R., *The Population Bomb*, New York: Ballantine, 1968.

Euromonitor International, 'Travel and tourism in Gambia', http://www.

euromonitor.com/Travel_and_Tourism_in_Gambia (accessed 11 May 2008).

Fackler, Martin, 'As oil prices rise, nations revive coal mining', *New York Times*, 22 May 2008, topics.nytimes.com/top/reference/timestopics/people/f/martin_fackler/index.html? (accessed 16 July 2008).

Fagan, Brian, *The Little Ice Age: How Climate Made History 1300–1850*, New York: Basic, 2000.

— *The Great Warming: Climate Change and the Rise and Fall of Civilizations*, New York: Bloomsbury, 2008.

Fairlie, Ian, 'Reasonable doubt', *New Scientist*, 26 April 2008, p. 19.

Fox, Douglas, 'Saved by the trees?', *New Scientist*, 20 October 2007, pp. 42–6.

Friedman, Milton, *Capitalism and Freedom*, 2nd edn, Chicago: University of Chicago Press, 1982.

Gaskill, Alvia, 'Global albedo enhancement project', *Library 4 Science*, http://www.global-warming-geo-engineering.org/Albedo-Enhancement/Introduction/ag1.html (accessed 27 May 2008).

Glover, Julian, 'They cheat, I tell you', *The Guardian*, 3 May 2008, p. 35.

Glynos, Jason and Howarth, David, *Logics of Critical Explanation in Social and Political Theory*, London: Routledge, 2007.

Gore, Al, *An Inconvenient Truth: The Movie*, Paramount DVD, 2006.

Gray, John, 'Those who control oil and water will control the world', *The Observer*, 30 March 2008, p. 33.

Hansen, Jim, 'A slippery slope: how much global warming constitutes "dangerous anthropogenic interference"?', *Climatic Change* (68:3) 2005, pp. 269–79.

Hardy, Thomas, *The Return of the Native* [1878], ed. George Woodcock, London: Penguin, 1978.

Hobbes, Thomas, *Leviathan, or The Matter, Forme, and Power of a Commonwealth Ecclesiasticall and Civill* [1651], ed. Richard Tuck, Cambridge: Cambridge University Press, 1991.

Howarth, David, 'Ethos, agonism and populism: William Connolly and the case for radical democracy', *British Journal of Politics and International Relations* (10:2) 2008, pp. 171–93.

Hutton, Will, *The Writing on the Wall: China and the West in the 21st Century*, 2nd edn, London: Abacus, 2008.

Keeley, Graham, 'Drought ignites Spain's "water war"', *The Observer*, 6 April 2008, p. 37.

Keenlyside, N. S. et al., 'Advancing decadal-scale climate prediction in the North Atlantic sector', *Nature* (453), 1 May 2008, pp. 84–7.

Klein, Naomi, *No Logo*, London: HarperCollins, 2001.

— *Fences and Windows: Dispatches from the Front Lines of the Globalization Debate*, London: Flamingo, 2002.

— *The Shock Doctrine: The Rise of Disaster Capitalism*, London: Allen Lane, 2007.

Krupp, Fred and Horn, Miriam, *Earth: The Sequel. The Race to Reinvent Energy and Stop Global Warming*, New York: W. W. Norton, 2008.

Laclau, Ernesto and Mouffe, Chantal, *Hegemony and Socialist Strategy: Towards a Radical Democratic Politics*, London: Verso, 1985.

Lambert, Richard, 'Fuelling the debate on climate change', *The Guardian*, Review Section, 19 April 2008, p. 9.

Lang, Tim, *Food Industrialisation and Food Power: Implications for Food Governance*, London: IIED, 2004.

Laurance, William, 'Expect the unexpected', *New Scientist*, 12 April 2008, p. 17.

Lawson, Nigel, *An Appeal to Unreason: A Cool Look at Global Warming*, London: Duckworth Overlook, 2008.

Lea, Ruth, 'Spare us the meddlers', *The Guardian*, 10 April 2008, p. 35.

Lewin, Roger, *Complexity: Life on the Edge of Chaos*, London: Phoenix, 1993.

Lindzen, Richard, 'Climate of fear: global-warming alarmists intimidate dissenting scientists into silence', *Wall Street Journal*, 12 April 2008, http://www.opinionjournal.com/extra/?id=110008220 (accessed 16 July 2008).

Lomborg, Bjorn, *The Skeptical Environmentalist: Measuring the Real State of the World*, Cambridge: Cambridge University Press, 2001.

— *Cool It: The Skeptical Environmentalist's Guide to Global Warming*, New York: Knopf, 2007.

Lovelock, James, *The Revenge of Gaia: Earth's Climate Crisis and the Fate of Humanity*, London: Allen Lane, 2006.

— 'The Earth is about to catch a morbid fever that may last as long as 100,000 years', *The Independent*, 16 January 2006, http://www.independent.co.uk/opinion/commentators/james-lovelock-the-earth-is-about-to-catch-a-morbid-fever-that-may-last-as-long-as-100000-years-523161.html (accessed 17 September 2008).

— and Rapley, Chris, 'Ocean pipes could help the Earth to cure itself', *Nature* (449), 27 September 2007, p. 403.

Lovett, Richard, 'Carbon lockdown', *New Scientist*, 3 May 2008, pp. 32–5.

Lunt, D. J. et al., '"Sunshade world": a fully coupled GCM evaluation of the climatic impacts of geoengineering', *Geophysical Research Letters* (35), 25 June 2008, L12710.

Lynas, Mark, *Six Degrees: Our Future on a Hotter Planet*, London: HarperCollins, 2007.

Lyotard, Jean-François, *The Differend: Phrases in Dispute*, trans. Georges Van Den Abbeele, Manchester: Manchester University Press, 1988.

Macalister, Terry, 'A change in the climate: credit crunch makes the bottom line the top issue', *The Guardian*, 6 March 2008, p. 28.

— 'Dawn of a new nuclear age', *The Guardian*, 22 March 2008, p. 47.

— 'Biofuels: a solution that became part of the problem', *The Guardian*, 25 March 2008, p. 11.

— 'Demands for crackdown on biofuels scam', *The Guardian*, 1 April 2008, p. 1.

— '"Dishonest, irresponsible": Shell lambasted for pulling out of world's biggest wind farm', *The Guardian*, 2 May 2008, p. 36.

McDougall, Dan, 'Time runs out for islanders on global warming's front line', *The Observer*, 30 March 2008, p. 46.

MacKenzie, Debora, 'The end of civilization', *New Scientist*, 5 April 2008, pp. 28–31.

— 'Are we doomed?', *New Scientist*, 5 April 2008, pp. 32–5.

McKibben, Bill, *Hope, Human and Wild: True Stories of Living Lightly on the Earth*, London: Little, Brown, 1995.

— *The End of Nature: Humanity, Climate Change and the Natural World*, 2nd edn, London: Bloomsbury, 2003.

Marglin, Stephen A., *The Dismal Science: How Thinking Like an Economist Undermines Community*, Cambridge, MA: Harvard University Press, 2008.

Meadows, Donella H. et al., *The Limits to Growth: A Report for the Club of Rome's Project on the Predicament of Mankind*, New York: Potomac Associates, 1972.

—, Meadows, Dennis L. and Randers, Jorgen, *Beyond the Limits: Global Collapse or a Sustainable Future*, London: Earthscan, 1992.

Monbiot, George, 'Are you paying to burn the rainforest?', *The Guardian*, 4 November 2007, p. 42.

— *Heat: How We Can Stop the Planet Burning*, 2nd edn, London: Penguin, 2007.

— 'Jobs are used to justify anything, but the numbers don't add up', *The Guardian*, 1 April 2008, p. 31.

— 'If there is a God, he's not green. Otherwise airships would take off', *The Guardian*, 6 May 2008, p. 25.

Moorhead, Joanna, 'Winds of change', *The Guardian* (Supplement, 'The green mile'), 30 August 2008, p. 1.

Mouffe, Chantal, *The Democratic Paradox*, London: Verso, 2000.

— *On the Political*, London: Routledge, 2005.

National Audit Office, 'UK greenhouse gas emissions: measurement and reporting', Spring 2008, http://www.nao.org.uk/publications/0708_greenhouse_gas_emissions.pdf (accessed 17 July 2008).

Overseas Development Institute,'The Gambia tourism value chain and its pro-poor impact', http://www.odi.org.uk/tourism/projects/06_gambiavca_html (accessed 11 May 2008).

Palmer, T. N. et al., 'Toward seamless prediction: calibration of climate change projections using seasonal forecasts', *Bulletin of the American Meteorological Society* (89:4) April 2008, pp. 459–70.

Pearce, Fred, 'Climate change: menace or myth?', *New Scientist*, 12 February 2005, pp. 38–43.

— *The Last Generation: How Nature Will Take Its Revenge For Climate Change*, London: Eden Project, 2006.

— 'Rising temperatures bring their own CO_2', *New Scientist*, 22 March 2008, p. 11.

— 'Saved?', *New Scientist*, 22 March 2008, pp. 36–9.

— 'Making a difference', *New Scientist*, 12 April 2008, pp. 50–3.

— 'Dirty, sexy money', *New Scientist*, 19 April 2008, pp. 38–41.

— 'Tracking the winds of climate change', *New Scientist*, 26 April 2008, pp. 48–9.

— 'We need better forecasts – and fast', *New Scientist*, 3 May 2008, pp. 8–9.

— *Confessions of an Eco-Sinner: Travels to Find Where My Stuff Comes From*, London: Eden Project, 2008.

— and Lomborg, Bjorn, 'An inconvenient voice' (interview with Bjorn Lomborg), *New Scientist*, 27 October 2007, pp. 54–5.

Pilkington, Ed, 'Climate target is guaranteed catastrophe', *The Guardian*, 7 April 2008, p. 1.

Ray, Dixy Lee and Guzzo, Lou, *Trashing the Planet: How Science Can Help Us Deal with Acid Rain, Depletion of the Ozone, and Nuclear Waste (Among Other Things)*, Washington, DC: Regnery Gateway, 1990.

Rice, Xan, 'Uganda "averts tragedy" with reversal of decision to clear virgin forest for biofuel', *The Guardian*, 29 October 2007, p. 22.

— 'Wildlife and livelihoods at risk in Kenyan wetlands biofuel project', *The Guardian*, 24 June 2008, p. 17.

Rosenzweig, Cynthia et al., 'Attributing physical and biological impacts to anthropogenic climate change', *Nature* (453), 15 May 2008, pp. 353–7.

Royal Commission on Environmental Pollution, 'The environmental effects of civil aircraft in flight', http://www.rcep.org.uk/aviation/av12-txt.pdf (accessed 17 July 2008).

Said, Edward, *Orientalism*, London: Routledge & Kegan Paul, 1978.

Sassoon, David, 'Breakthrough: concentrated solar power all over southwest US', http://solveclimate.com/blog/20080117/breakthrough-concentrated-solar-power-all-over-southwest-us, 17 January 2008 (accessed 22 May 2008).

Scheffer, Martin, Brovkin, Viktor and Cox, Peter M., 'Positive feedback between global warming and atmospheric CO_2 concentration', *Geophysical Research Letters* (33), 26 May 2006, L10702.

Schirber, Michael, 'Ice ages blamed on tilted Earth', http://www.livescience.com/environment/050330_earth_tilt.html, 30 March 2005 (accessed 8 May 2008).

Sim, Stuart, *Fundamentalist World: The New Dark Age of Dogma*, Cambridge: Icon, 2004.

— *Empires of Belief: Why We Need More Scepticism and Doubt in the Twenty-First Century*, Edinburgh: Edinburgh University Press, 2006.

Simon, Julian L., *The Ultimate Resource*, Princeton, NJ: Princeton University Press, 1981.

Sington, David, 'Global dimming', http://www.bbc.co.uk/sn/tvradio/

programmes/horizon/dimming_prog_summary.shtml (accessed 25 April 2008).

Sloman, Lynn, *Car Sick: Solutions for Our Car-addicted Culture*, Totnes, Devon: Green Books, 2006.

Smith, Dan and Vivekananda, Janani, *A Climate of Conflict: The Links Between Climate Change, Peace and War*, London: International Alert, 2007.

Snyder, Rachel Louise, *Fugitive Denim: A Moving Story of People and Pants in the Borderless World of Global Trade*, New York: W. W. Norton, 2008.

Sparrow, Andrew and Wintour, Patrick, 'Nuclear is UK's new North Sea oil – minister', *The Guardian*, 26 March 2008, p. 1.

Stern, Nicholas, *The Economics of Climate Change: The Stern Review*, Cambridge: Cambridge University Press, 2007.

Stiglitz, Joseph, *Globalization and its Discontents*, London: Penguin, 2002.

— *Making Globalization Work*, London: Penguin, 2006.

Strange, Susan, *Casino Capitalism*, Oxford: Blackwell, 1986.

Stuckler, David, King, Lawrence P. and Basu, Sanjay, 'International Monetary Fund programs and tuberculosis outcomes in post-communist countries', *PLoS Medicine* (5:7), July 2008, DOI: 10.1371/journal.pmed.0050143.

'Sunshield to cool planet would decimate ozone layer', *New Scientist*, 3 May 2008, p. 17.

Tainter, Joseph A., *The Collapse of Complex Societies*, Cambridge: Cambridge University Press, 1988.

Titus, James G. et al., 'Greenhouse effect and sea level rise: the cost of holding back the sea', http://www.epa.gov/climatechange/effects/downloads/cost_of_holding.pdf.

Tonder, Lars and Thomassen, Lasse, eds, *Radical Democracy: Politics Between Abundance and Lack*, Manchester: Manchester University Press, 2005.

'An unsustainable scam', *The Guardian*, 1 April 2008, p. 34.

US National Oceanic and Atmospheric Administration, 'Global warming: frequently asked questions', http://www.ncdc.noaa.gov/oa/climate/globalwarming/html (accessed 17 July 2008).

Vidal, John, 'The looming food crisis', *The Guardian*, G2 Section, 29 August 2007, pp. 5–9.

— 'Government figures hide scale of CO_2 emissions, says report', *The Guardian*, 17 March 2008, p. 1.

— 'Britain seeks loophole in EU green energy targets', *The Guardian*, 29 March 2008, p. 1.

VisitBritain, http://www.tourismtrade.org.uk/MarketIntelligenceResearch/KeyTourismFacts.asp (accessed 6 August 2008).

Walker, Gabrielle, *Snowball Earth: The Story of the Great Global Catastrophe that Spawned Life As We Know It*, London: Bloomsbury, 2003.

— and King, Sir David, *The Hot Topic: How to Tackle Global Warming and Still Keep the Lights On*, London: Bloomsbury, 2008.

'Warming oceans starved of oxygen', *New Scientist*, 10 May 2008, p. 17.

'We could use moon as a giant flashlight', *New Scientist*, 26 April 2008, p. 17.

'We'd like another half a planet please', *New Scientist*, 3 November 2007, p. 13.

Wolf, Martin, *Why Globalization Works*, New Haven, CT: Yale University Press, 2004.

'The world's coasts under threat', *New Scientist*, 1 September 2007, p. 10.

World Travel & Tourism Council, http://www.wttc.org/eng/Tourism_News/Press_Releases/Press_Releases_2008/Continued_growth_signalled_for_Travel_and_Tourism_Industry/ (accessed 23 June 2008).

http://www.wttc.org/eng/Tourism_Research/Tourism_Satellite_Accounting/TSA_Country–Reports/France/ (accessed 23 June 2008).

http://www.wttc.org/eng/Tourism_Research/Tourism_Satellite_Accounting/TSA_Country_Reports/Greece/ (accessed 23 June 2008).

http://www.wttc.org/eng/Tourism_Research/Tourism_Satellite_Accounting/TSA_Country_Reports/Peru/ (accessed 24 June 2008).

Worldwatch Institute and Center for American Progress, 'American energy: the renewable path to energy security', http://americanenergynow.netAmericanEnergy.pdf (accessed 22 May 2008).

Wright, Lawrence, 'The rebellion within', *The Observer Magazine*, 13 July 2008, pp. 14–49.

Zeng, Nin, 'Carbon sequestration via wood burial', *Carbon Balance and Management* (3:1), 3 January 2008 (http://cbmjournal.com/content/3/1/1).

Zweibel, Ken, Mason, James and Fthenakis, Vasilis, 'A solar grand plan', *Scientific American*, January 2008, pp. 64–73.

Index